THE PERFECT SPECIMEN:
THE 20TH CENTURY RENOWN BOTANICAL COLLECTOR--
YNES MEXIA

by Durlynn Anema, Ph. D.

The Perfect Specimen: The 20th Century Renown Botanical Collector--Ynes Mexia by Durlynn Anema, Ph. D.
Copyright © 2019 by Durlynn Anema, Ph. D.

All rights reserved. No part of this book may be reproduced or transmitted in any form or by any means, electronic or mechanical, including photocopying, recording or by any information storage and retrieval system, without permission from the author, except for inclusion of brief quotations in a review.

ISBN 13: 978-0-88100-170-9
ISBN10: 0-88100-170-8
Library of Congress Number: 2019946615

Cover Design by NZ Graphics
Cover Photograph Courtesy of the California Academy of Sciences

Published by National Writers Press

Library of Congress Cataloging-in-Publication Data

Anema, Durlynn
The Perfect Specimen: The 20th Century Renown Botanical Collector--Ynes Mexia by Durlynn Anema, Ph. D.

International Standard Book Number 10: 978-0-88100-170-9
International Standard Book Number 13: 0-88100-170-8

1. History/Expeditions & Discoveries
2. Biography & Autobiography/ Adventurers and Explorers
3. History/ Modern/ 20th Century   I. Title   2019946615

10   9 8 7 6 5 4 3 2 1

Dedicated to those people who gained confidence in their abilities later in life.

ACKNOWLEDGEMENTS

Ynes Mexia's story is compelling because it demonstrates no matter your age or circumstances you can accomplish whatever you want. Mexia exemplifies the power of exploration, discovery, faith, and positive attitude. An additional factor is Dr. Philip King Brown who demonstrates that reliance on yourself rather than your doctor is the secret to healing. My gratitude to Kate Davis (wherever you are) for bringing Mexia into my life. And to Morgan Reynolds Publishers for believing in my Young Adult story about Mexia, my first book about her.

No book is written in isolation, especially a biography. Research is essential, and I am forever in debt to those people who provide the research facilities. My original research took me to Michele Wellck, retired Academy Archivist, California Academy of Sciences, San Francisco, who took personal charge of my project. Others who helped were Denise J. Price, Save-the-Redwoods League; Caitlin Lewis, assistant librarian, William E. Colby Memorial Library, Sierra Club, San Francisco; and the patient 2004 personnel of the Bancroft Library, University of California, Berkeley. The Bancroft Library is the repository for the vast majority of Mexia's correspondence, which represents most of the documentation of her life.

For this adult version I thank Jennifer Verhines, Donor Relations Manager, Save-the-Redwoods League and the League's excellent photographers; Seth Cotterill and Rebekah Kim, Archivists, California Academy of Sciences, San Francisco; ShienDee Pullman, Library Director, Gibbs Memorial Library, Mexia, Texas; Archives Section, Texas State Historical Association; and Charlotte Mangin who has made Mexia part of her PBS salute to early twentieth century American heroines. *Unladylike2020* can be viewed March 2020.

Durlynn Anema, Ph.D.

Finally, deep appreciation to National Writers Press Publisher, Anita Biewenga for always believing in the power of my books.

Table of Contents

Author's Note 8
Preface 9
Chapter One 11
Chapter Two 17
Chapter Three 27
Chapter Four 37
Chapter Five 43
Chapter Six 47
Chapter Seven 55
Chapter Eight 71
Chapter Nine 79
Chapter Ten 87
Chapter Eleven 99
Chapter Twelve 111
Chapter Thirteen 119
Chapter Fourteen 131
Chapter Fifteen 137
Timeline 144
Mexia's Expeditions 145
Additional Chapter Information 147
Bibliography 165

Author's Note

Most successful people, no matter their endeavor or occupation, find inspiration through either a parent, an important or inspirational person, or an event. When I wrote about female explorers Harriet Chalmers Adams and Louise Arner Boyd this was the case. Both had strong fathers who realized they never would have sons (Boyd's two sons had died) and transferred their aspirations to their daughters who became strong, confident women.

This is not the case with Ynes Mexia. A shy, quiet girl, she seemed to fade into the background with both her parents. She led a lonely life, which ironically aided her in her later endeavors.

Mexia's is a story of retreat into self in early years, then a blossoming to reach her highest potential after fifty-years-old. It also is the story of a doctor who, during the infancy of psychiatry and psychology, mentored this woman to her potential and became the father figure she never had.

Read, marvel, and enjoy Ynes Mexia's story.

Preface

"I don't think there's any place in the world where a woman can't venture alone. In all my travels I've never been attacked by wild animals, lost my way or caught a disease," Ynes Mexia told a *San Francisco News* reporter in 1937.

Ynes Mexia never dreamed that one day she would explore North and South America as a botanical collector, obtaining between 140,000 and 150,000 specimens, more than any other woman collector of the era. Her childhood had few adventures, and a loneliness that resonated throughout her adult years. She moved often during her youth, usually attending boarding schools, and finding it difficult to interact with either her peers or adults. As a young adult she encountered situations she could not understand, so retreated further into herself. Yet these challenges enabled her to adapt to the solitary nature of her work in some of the most remote areas of the American continents.

While these early consequences led to serious complications as Mexia matured, they ultimately brought her to a place where she overcame her reticence and discovered her abilities. Always enthralled by nature, her transition to botany was a logical progression. She used her love of solitude to her advantage, traveling alone accompanied only by local guides, carrying only what she could manage to sustain herself and support her work in gathering specimens. From a world she couldn't understand, she moved into a world of plants native to the Americas and produced an archive of new records and species which still is studied today.

Mexia was the most accomplished female plant collector of her time.

Chapter One

A Lonely Life

When researching creative and over achieving individuals, it often is assumed they were influenced by parents and/or their family situation. This usually is the case, either father or mother as the motivating factor, often both. Therefore, Ynes Mexia does not fit this mold in any way, having neither encouraging parents nor loving ones. Hers was a lonely life, with only one place where she felt secure -- the Catholic Church. She attended Mass regularly wherever she was in the civilized world, feeling that only in a sacred cathedral or church could she feel love -- that of Jesus Christ.

Yet, in spite of never feeling any love (or very little) from her family, of moving often from town to town or school to school, of never quite fitting into her school or social environment, Mexia definitely became an overachiever. These challenges enabled her to adapt to the solitary nature of her work in some of the most remote areas of the American continents.

Born on May 24, 1870, in the Georgetown section of Washington, D. C., Mexia lived only a short time in that city. During that short year she was baptized Ynes Enriquetta Julietta Mexia, probably by Samuel Eccleston, the Roman Catholic archbishop of Baltimore, who was related to her mother. She remained a strong Catholic throughout her life, always trying to attend Mass when a church was available.

Durlynn Anema, Ph.D.

Mexia's parents were Enrique and Sarah R. Wilmer Mexia who married in 1868. Their first child was a daughter, Adele. Both had been married previously. They met in Washington, D. C. while Mexia's father was serving in the American capital as a representative of the Mexican government under President Benito Juarez.

Sarah, raised in Baltimore in pampered luxury, had two children from a previous marriage that ended in divorce. Her brother George Louis Hammeken was involved with the building of a Mexican railroad in the 1850's, and was married to Enrique's sister Maria Adelaida Matilda Mexia. Perhaps this is how Enrique and Sarah met.

Enrique was born in Mexico City to Jose Antonio Mexia and Charlotte Walker de Mexia in January 1829. His mother was from Southampton, England. His father was a Mexican general under Mexican President Santa Anna and a leader in Mexico's Federalist Party. (See Additional Chapter Sources for description of Jose Antonio Mexia.)

On May 3, 1839, when Enrique was ten and living in New Orleans with his older sister and mother, his father was executed by General Antonio Lopez de Santa Anna as a co-conspirator in an uprising against the Mexican government. The family returned to Mexico City, then moved back to New Orleans in 1844 so Enrique and his sister could be educated in the United States. Thus, Enrique found himself regularly moving between the U. S. and Mexico throughout his early life.

Enrique wanted to imitate his father, so he joined the Mexican military, earning the rank of brigadier general. He fought against the French during Maximilian's reign in Mexico and then served in the Mexican Legation in Washington, D. C.

In addition to his position with the Mexican government, Enrique also worked with Hammeken as a land agent in Texas. Texas land was part of his and his sister's inheritance when his father was executed. In November 1833, before Texas was admitted to the union, the state of Coahuila and Texas granted Jose Mexia's children separate titles to eleven-league tracts of land in Limestone, Freestone, and Anderson counties. Enrique served as the land agent, overseeing the family's large

THE PERFECT SPECIMEN

landholdings in the three counties. He also held a grant of six tracts of land on the Trinity River in Freestone and Anderson Counties, Texas.

In 1870, the Houston and Texas Central Townsite Company purchased 1,920 acres of the Mexia grant to build a town. Enrique was credited with arranging the sale and also donated land for the town. In 1871, the Houston and Texas Central Townsite Company named the town Mexia in honor of the Mexia family.

When Ynes was a year old, her family moved to Mexia, Texas in what is now Limestone County. It is located directly east of Waco and south of Dallas. The family built a home on their Limestone property.

The City of Mexia, Texas

Depending on the reference, the city was named either for Jose Antonio Mexia or for Enrique. The area is near where the rolling hills of the great plains begin. These hills provided grazing land for the buffalo herds. Consequently, the Comanche tribe came into conflict with white settlers in and around this area.

The town was laid out in 1870 by a trustee of the Houston and Texas Central Townsite Company. They offered lots for sale in 1871 when the Houston and Texas Central Railway was completed between Hearne and Groesbeck. The Mexia post office began operation in 1872, and the community was incorporated in 1873 by an act of the Texas State Legislature. By 1880, Mexia had four schools, three churches, and a variety of businesses to serve its one thousand eight hundred residents.

This was during the time Ynes was in Mexia as a young child, so she saw the town grow.

The town's biggest boom happened in 1912 when a large natural gas deposit was discovered. Then in 1920 the oilfield was discovered and the population increased from 3,482 to nearly 35,000. This prosperity continued until the 1930's. In 2000, the population was 6,500 people.

The Mexia House built in 1871 by Enrique Mexia for his family when they moved to Mexia, Texas.
Courtesy of *Mexia Daily News,* December 9, 1992.

The Mexia, Texas Town Opera House (with arrow pointing toward it), was built in 1869 and destroyed by an explosion is 1916. Photo appeared in the *Mexia Daily News.*

The Perfect Specimen

As Ynes grew up, she kept to herself and had few friends. Her mother Sarah and older sister Adele were busy with social engagements and entertaining friends. These events did not interest Ynes, who preferred the outdoors where she could experience nature and its wonders.

Her father Enrique often was away on business, so she hardly knew him. When he was at home he paid little attention to either girl, obviously wishing he had a son to carry on his military career.

Consequently, Ynes was largely ignored. Because she was shy and quiet, this did not bother her. She preferred to read or walk in the fascinating countryside. She watched the birds and small animals, and examined the flowering plants. When she was six years old, she began school, which she liked very much. However, because of her shyness she had few friends. She could not understand how to get along with her peers and retreated further into her books.

In late 1879, Ynes' small world changed. Her parents separated. Her father headed to Mexico City and his property in that country. He often traveled to Mexia, Texas to oversee his holdings but never contacted Ynes. He evidently was quite active in Mexico City because he is credited with installing the first electric lights there.

Ynes, her mother, and her sister went to Philadelphia, Pennsylvania. Her mother enrolled her in a private school where she felt like an outsider. The girls in the school were more sophisticated, having grown up in the city. She could not understand their ways, nor they hers. Consequently, she retreated more within herself, not only feeling unloved by both parents but also by everyone else around her. Her sister Adele spent little time with Ynes, preferring socialite friends.

Ynes' one solace was when she was in church or walking the church grounds. She loved to kneel in front of the Virgin Mary to pray and tell her about her life. She felt this kindly woman, who was so beautiful, might be able to understand her. (Probably, only this statue saw that Ynes cried to be loved.)

Durlynn Anema, Ph.D.

Her schooling continued at boarding schools in Toronto, Canada and St. Joseph's College, Emmitsburg, Maryland. Each move deepened her loneliness. Beside finding solace in reading and exploring the environmental world, she also wrote copious letters to her father in Mexico. Desperately, she hoped that at some point he would visit her as in this letter on January 14, 1882: "To Papa: When are you coming? Going to dancing school."

Her letters contained vivid descriptions of her surroundings, demonstrating her close attention to the world around her as well as her careful eye for detail. These traits later helped her in botanical collecting and she often received accolades for her detailed botanical descriptions.

When Ynes finished St. Joseph's College she received a letter which both excited and terrified her -- excited to once again see her father, and terrified of what was going to happen to her. Her father sent for her to come and live with him in Mexico City.

While this would be a completely new experience in a country where she had never been, she knew she had to obey her father, and was determined to do her best.

Chapter Two

Adjusting to a New Life

Ynes obeyed her father's request although she knew nothing about Mexico or the father she had not seen for years. Obeying a parent was what she had learned, even if that parent had shown neither love nor caring of her. As she traveled south on the train to meet her father, she had no idea of the different life she was about to enter.

Her father did not meet the train, sending one of his ranch hands to greet her and take care of her luggage. They drove in a horse and buggy to her father's hacienda on his ranch just outside Mexico City. As Ynes stepped off the train she realized she did not know one word of Spanish, yet that was the only language she heard. While she was frightened and worried, she also was determined to do whatever was asked of her. Maybe her father would be glad to see her and perhaps even love her.

Mexia was awed by the handsome man who greeted her at the sprawling hacienda. As she walked into the cool entry she stepped into a new world. Immediately, she loved the dark beam ceilings and red tile floors. After being shown to her room to refresh, she emerged and went to her father's study. There he told her what would be expected of her -- to supervise his entire household! It was a huge household -- cooks, maids, gardeners plus workers on the ranch itself.

Enrique explained that he would be away often because he was an official in the Diaz government of Mexico. She not only was to supervise the household, but also to plan dinners and banquets for any distinguished guests he might bring home. Her father expected her to succeed, unconcerned that she may be uncomfortable with her new role. Quickly, she realized he was cold and uncaring, using her only to manage the house and conduct ranching business. However, Ynes already had coped with schools she didn't like and classmates who ignored her, so this was just another obstacle she had to overcome.

Ynes was a loyal daughter, though she and her father were not close. The only time she saw him was across the table at dinner -- and even those times were rare. The biggest shock for this cultured girl, who knew very little about life, was the way her father lived. She heartily disapproved of her father's succession of mistresses. She never knew what woman she might see in their home -- or how long the woman would stay. And she couldn't comprehend what was going on between her father and these women, having no knowledge of the sexual side of a male/female relationship.

Imagine her surprise to discover she had a half sister, Amanda Gray Mexia (nicknamed Amy) and a half brother, Clarence W. Mexia. They were children of Mary Gray Mexia who Enrique married in 1867. They also lived in the hacienda and expected Ynes to wait on them as the servants did. She didn't like the way her new brother acted toward her, cold and distant, so never was close to him. However, she and Amy eventually developed a comfortable relationship.

Slight of stature with luxuriant brunette hair and sparkling brown eyes, Mexia was withdrawn and unacquainted with young men her age. According to the custom of the time, she rarely was alone with any man and thus had little interest in courtship. Her duty was to her father so she thought nothing about men or possible marriage. In fact, she had decided that when her father died she would become a nun.

Enrique became ill during the summer of 1896. During this time Ynes was on the ranch while her half-sister Amy was in Mexico City with their father. Evidently, by this time there was

THE PERFECT SPECIMEN

animosity between Ynes and Amy. Ynes tried to stay in touch to find out how her father was and Amy responded in this way:

August 5, 1896: "Getting better." (a typed letter)
August 7, 1896: "Papa better."
August 11, 1896: "Papa worse."
August 12, 1896: "Papa may die in a few days and don't come. He knows no one."
August 16, 1896: "Papa better, then worse."
August 17, 1896: "Remember that Papa dead or alive, into this house you cannot come."

Imagine how Ynes felt at that moment. She had wanted to be by her father's bedside at the end, yet received that threatening letter from Amy. Once again family had deserted her, did not want her around. Consequently, she could only wait for his death while at the ranch -- and wonder what the future would hold.

Enrique died on September 19, 1896 leaving a large estate both in Mexico and in Texas. Adele and Ynes were the sole beneficiaries in his Will with the exception of some small bequests, and were named as execurtixes of the Will. The Will also stipulated that if Ynes became a nun she would be cut out of the inheritance she shared with her sister. Ynes learned that Enrique had not formally married the mother of Clarence and Amy, which presented a problem.

The settling of the Will proved to be a long fight for Adele and Ynes. The Will was contested in probate court by Clarence Mexia and Miss Amada Gray Mexia who claimed they were legitimate children and heirs.

On July 26, 1899, Attorneys Harris, Etheridge and Knight, who represented Ynes and Adele in the probate fight, wrote to Messrs. Kountze Bros., Bankers, New York City. They also were associated with Maj. L. J. Farrar, Enrique's personal attorney for more than twenty years prior to his death and also his confidential friend and adviser.

"This will is contested in the probate court of this county by Clarence Mexia, a natural son of Gen'l Mexia, to whom he left a small property, and by the natural daughter of Gen'l Mexia, Miss Amada Gray Mexia, so-called, the daughter of Mrs. Mary Gray Mexia, so-called. The contestants claim that Gen'l Mexia was at the time of the execution of his last will of unsound mind and incapable of executing the will, and the contestant Amada Gray Mexia, so-called, claiming in addition that she is the daughter of Gen'l Mexia by Mary Gray Mexia, so called, with whom he had a common law marriage in the state of New York or New Jersey in the year 1867; and that she is the only issue of that marriage; and that Gen'l Mexia's marriage to Mrs. Sarah R. Ramsey in Patterson, New Jersey, on Sept. 5, 1868 (which is duly shown by the church and legal records of said city) was void, and that Miss Adele and Mrs. Ynes Mexia de Laue (Ynes had married in 1897) the fruit of that marriage are, in consequence, illegitimate."

In a later paragraph the letter says: "We believe from evidence in our possession that Gen'l Mexia was not in New York in 1867, and not until some time in July, 1868. The proof seems to show that he was in the Mexican Army in active service in 1867, and that he could not have been in New York that year."

The Perfect Specimen

These evidently were excellent attorneys because Adele and Ynes finally won and became the sole heirs of Enrique's estate. Adele never did go to Mexico or claim her share there. Ynes now became not only the manager of the ranch, but the proprietor. Because she had managed it for so long this was an easy transition.

Although Ynes had not been courted during earlier years, after her father died she suddenly found herself in that position. Whether it was because of her appearance, which was pleasantly attractive, or because of her inheritance could be debated. However, one man, Herman Laue, a young German-Spanish merchant, persisted in pursuing her. She was overwhelmed by his attention, and pleased to find she now was like other young women she knew -- a woman attractive to men. When he proposed shortly after her father died, she accepted having no idea of what marriage entailed. They were married in 1898 in a quiet ceremony and went to live in Tacubaya, near Mexico City, on a ranch she had inherited.

Laue managed the ranch and tried to take care of Ynes, a role she found hard to accept. She could not have a close relationship with her husband, an attitude developed during an entire lifetime of being by herself. She was not capable of understanding how to meet another person's intimate needs because hers had never been met. She could not grasp the concept of love because it had not become part of her nature -- that two people could become one both physically and mentally. She had no real understanding of communication with another human, especially a man. Her father had been distant, which led her to believe all men were the same way. In her own way, she tried to make Laue as happy as she could because she wanted the marriage to succeed. On his side, he may have compensated with other companions, as was the way for many Latin men. Then Laue died abruptly in 1904 after six years of marriage.

Now Mexia was truly alone in the world. Her mother and sister Adele were in the United States, so she had no contact with them. Her half sister Amy rarely visited, especially after the dispute over the Will. She remained on the ranch, taking over managerial duties, running the business, and becoming an astute financial entrepreneur, a skill she retained over a lifetime.

In 1908, she met and married Augustin de Reygadas, who was sixteen years younger. Like any marriage of this type, one must only speculate why he married Ynes. He truly could have loved her, or he was interested in her money and the ranch, or he needed a "mother" figure.

Whatever the reasons, Mexia tried to make the marriage work. However, at this stage of her life she had retreated too far within herself. She could give little affection and definitely not the intimacy expected in a marriage.

Mexia's spiraling unhappiness had severe physical consequences. She would often stay in her room curled into a fetal position on her bed. Her husband was beside himself, unable to understand her or what was happening. He took her to doctors and hired nurses, but nothing helped.

In 1909, Mexia suffered a mental and physical breakdown so severe her physician advised her to leave Mexico and find a competent doctor in the United States. Augustin agreed, aware her physical problems were a result of mental ones. She needed the most competent treatment available which he realized was in the United States. Her physician referred her to Philip King Brown, a noted doctor in San Francisco and Berkeley, California.

Mexia and her husband lived briefly in a hotel before finding an apartment in San Francisco. Dr. Brown quickly understood her condition and started a treatment which essentially was psychiatric. As he worked with Mexia, she began a long, slow movement toward recovery. Each day Reygadas hoped she might improve so they could return to Mexico. He felt uncomfortable in the U. S. because it was a culture quite different from his own and because he spoke little English. However, Dr. Brown constantly emphasized progress would be slow.

Mexia understood; de Reygadas did not. As they continued to live in San Francisco, she became concerned about the ranch, worried that if no one was overseeing it she would lose it. Among the enterprises she had in Mexico was a poultry business

THE PERFECT SPECIMEN

she had started at her father's hacienda outside Mexico City. She was especially concerned about it. As her concerns escalated, she insisted Reygadas return to Mexico to discover what was happening.

Quietly and convincingly, Mexia would say to Reygadas, "We can't let the ranch and businesses go. You have to return to Mexico."

Reygadas listened, knowing she was right but reluctant to leave. Finally, he nodded his head, "I will return." He left in mid-1910.

During the next ten years, the couple's communication was through their detailed correspondence. She called him "Petsito" and he called her "Petsita." He often begged to come to the United States but Mexia would point out that he knew little English and had no skills with which to make a living. He realized she was right and unhappily agreed to stay in Mexico.

These samples of Mexia's early letters also demonstrate her progress in healing as Dr. Brown worked with her.

November 20, 1910: Had a hat redone and looks 'better than new.' Having clothes altered, buying new coat. (Indicates she again was beginning to have an interest in her appearance.)

November 25, 1910: "Sell Hypothicary Bonds as decided 'a long time ago.' " (Evidently he was paying off his debts.)

"Glad Revolution in Mexico did not amount to much this time, but after President Diaz dies there will be a more serious one."

She also answered a letter Reygadas sent telling her he had decided to invest part of her capital in the United States, part in Mexico. She replied: "You are becoming more business like." Your Petsita

November 27, 1910: To Petsito: "So much business matters -- better now, such terrible state of confusion since I became sick."

November 30, 1910: Petsito: "Sorry so few sales. Have you tried advertising? God bless you and make you a good boy." Much love from your Petsita.

> December 27, 1910: "Went to Communion on Christmas morning and did not forget you."
>
> January 4, 1911: (Augustin had written and said he was lonely and wanted to join her.) She reminded him he went home to work on the house so it could be sold.
>
> To her Banker on January 18, 1911: Worried about a loan she had and, because she wanted no debts, she wanted to pay it off.
>
> January 28, 1911: Sent more money to Augustin.
>
> February 26, 1911: "Well and happy. Making baskets." (This has to do with her treatment as does the next letter.)
>
> March 11, 1911: Visiting poor people. Distributing clothes. "becoming quite vain of my skill as a basket maker."
>
> March 22, 1911: She told him he must sell house. "You have not been made a martyr of."
>
> March 29, 1911: He still wanted to come but she said she was planning a reunion with her sister Adele in the East so he couldn't come. Then she reminded him he had not sold the house. She said he could come when she returned.
>
> March 31, 1911: Evidently the house sold and she urged him to stay in Mexico to be sure to sign all the papers.
>
> April 11, 1911: She was not sending him any money and he was urged to stay on the ranch.
>
> June 15, 1911: She was still paying bills.

As Mexia's health improved she realized that her marriage was impossible. On July 31, 1911 she wrote a letter to Reygadas in which she bluntly told him the situation, that they could not continue their marriage. In her new state of mind she had decided to make some decisions to move her life ahead. One of these was admitting she could no longer be married. The entire letter, which is in Additional Chapter Sources, shows her reasoning in reaching this decision.

She starts with her usual salutation: "Dear Petsito;"

The Perfect Specimen

Then she talks about his letter which had just arrived. "Your letter of July 19th has reached me, and I think you have taken a very wrong and mistaken idea of what I wrote to you about coming." She goes into details about what his life would be in the U. S., saying "it did not seem fair to me to let you come without some idea of what your circumstances would be ..."

First, she mentions he would have to work and make his own living. Second, she says he couldn't bring any plants or animals because he would have to live in a boarding house.

The third point is the basis of the letter, that she was "unable to live with you as your wife, and I do not impose this on you as a condition on which I will allow you to come here, as you seem to think, for if you come here this will have to be exactly the same." She is straight forward in saying it had nothing to do with him, nor was she trying to punish him. Rather it was her "inability to bear with the marriage relation ... I only wish that this were not so, but see no way to remedy the situation."

Commenting further, she talks about how ill she was when they married and that the nervous disturbances she suffered were due to the marriage relation, yet he showed no consideration of her. With their separation, she had improved.

"I have asked of you no sacrifices that I have not made myself, and except this separation from me, I think you cannot complain; you have been in the country you prefer, have had your family and your sister to live with you, have had your animals and other things to your liking, and have been provided for..." After that she added that she saw no necessity for him to come.

Evidently, she told him before they were married that she had a great fear of the marriage relationship, and freely admitted she did not have a "very affectionate disposition...I think it is more honest and fair to you to say this clearly, but what I can also say is that I take a very real interest in you and your happiness and well being..."

(She signed it) Yours affectionately,
Pesita

Reluctantly, Reygadas accepted the fact he would not return and went on with his life. They remained separated for the remainder of their lives with sources varying on whether they ultimately divorced.

She continued writing Reygadas for the rest of the year.

October 24, 1911: She doubted she should keep the business in his hands. "You were so enthusiastic about running the business successfully that against my better judgement I allowed you to carry it on at a great loss to myself for two years."

November 27, 1911: 'I have come to the end of my resources and I must stop this business that is ruining me." She urged him to leave the house and go to work.

December 29, 1911: "I hope by this time you have found a suitable position."

Their letter writing continued for several years, this time only as friends. Once Mexia had removed Reygadas from her life, she could concentrate on her treatments by Dr. Brown. She realized he had become the mentor and father figure she had always needed, thrilled she finally was beginning to understand herself.

CHAPTER THREE

FINDING A NICHE

Dr. Philip King Brown became more than a doctor treating Mexia's physical, emotional, and mental problems. He became her friend, mentor, and the father figure she never had, and their friendship remained during her entire lifetime.

Dr. Brown had a unique method of treating his psychiatric patients which aided Mexia's healing progress. At first, he saw her often to treat her anxiety but was careful from the beginning not to have her rely on him. He would tell her to call him only when she felt she needed him and to start to rely again on her strengths. Conscientiously, she followed all orders he gave.

The study of psychiatry and psychology was only fifty years old when Mexia arrived in San Francisco. Little was known about using prescription medications to treat mental illnesses, nor were physicians and psychiatrists clear on how mental conditions could affect a person's physical health. Dr. Brown had studied the effects of stress on mental functions and decided to use this new found knowledge to move Mexia toward recovery.

His wide range of abilities led him in several directions. One of these was the concept of having patients become involved with a hobby or work to move them beyond thinking about their condition. In 1911, he established Arequipa Sanatorium in Fairfax north of San Francisco. Its main purpose was tuberculosis treatment. Part of the treatment for both in and out patients was to have them become involved in some type of hobby. The sanatorium, which had pottery making facilities as part of the therapy, became known as Arequipa Pottery from 1911 through 1918.

While Mexia did not make pottery, she did become interested in basket making. As mentioned in her letters to Reygadas, she made several baskets, gaining confidence in her skills.

Dr. Philip King Brown

Dr. Brown was a highly regarded physician throughout the United States. He graduated from Harvard Medical School in 1893, then became an assistant in nervous disease at the University of California Medical School in 1894. He taught at Medical Schools and practiced at hospitals throughout the Bay Area including Stanford University Hospital. Interested in all phases of mental and physical conditions, he was the perfect doctor for Mexia when she arrived in San Francisco.

Dr. Brown was born in Napa, California in 1869. His mother, Dr. Charlotte Blake Brown, was founder of San Francisco's Children's Hospital, and an outstanding physician and surgeon. In 1909, Brown married Helen Hillyer, and the couple had four children. He was co-founder of the San Francisco Boys Club, and active in the Tuberculosis Polyclinic, designed to help people recognize the symptoms of tuberculosis and to cure themselves when they contracted the disease.

Although Dr. Brown considered himself a general practitioner, he was well-known for his work with tuberculosis at the Polyclinic. He was incensed by the attitude of most municipal authorities toward the treatment of tuberculosis which led him to found the Arequipa Sanatorium, financed and built almost entirely by donations.

THE PERFECT SPECIMEN

Except for a brief stint with the Red Cross in France during World War I, Dr. Brown continued as Medical Director of Arequipa until the early 1930's. He died in October 1940, having remained active in many charities and worthy causes in the Bay Area.

Brown's concern for women's health -- remarkable for his time -- can be credited in part to his mother's influence. Charlotte Blake Brown was a leader in women's and children's health, and one of San Francisco's first female physicians. Brown concluded that the TB rate among working-class women was twice that of men, attributing this difference to the environment in which women's roles placed them (indoors versus outdoors).

Not only were working women particularly vulnerable to infection, Brown noted, but they also were uniquely deprived of medical attention. Because women's work continued inside the home, they never had the opportunity to rest and recuperate. They also could not afford sanatorium care.

Holism, humanism, and a keen sense of what would today be called "occupational justice" distinguished Brown's professional endeavors. Dr. Brown directed energy toward the threat of TB and a medically deprived population. Before founding Arequipa, Brown directed the San Francisco Polyclinic for several years, trying the same approach that later characterized Arequipa. Close observation of his patients convinced him of the connectedness between person, occupation, and environment. "Every effort is made to study all the physical, social and financial aspects of each patient." (See Additional Chapter Sources about Arequipa Sanatorium.)

Durlynn Anema, Ph.D.

Dr. Brown showed an interest in all facets of Mexia's activities, something she had never experienced. As he became the father figure she never had, she confided to him that she desperately wanted peace of mind. He assured her that it was possible.

For the first time in her life she had a person who believed in her. He quietly but continually helped her see the positive things she had done during her lifetime, until she began to feel more self-confident. He suggested physical activities including walking, which she always had enjoyed. And when he suggested she might want to make baskets she decided she would try -- assured enough to feel she could do a new task.

Mexia enjoyed living in San Francisco, with its soaring hills hugging the bay and the churning Pacific Ocean, and gained strength by walking its neighborhoods -- and, soon up and down the hills. The city slowly was recovering from the 1906 earthquake, reestablishing itself as the bustling financial and cultural center of the West. She was amazed at how fast the city had recovered. It seemed to her that all the residents wanted to make a miracle occur immediately -- a miracle she saw happening right before her eyes. The rubble from the quake and fire had been cleared and new buildings were rising. The transportation system -- street cars and cable cars -- had been restored. She realized that when people wanted to succeed they could in remarkable fashion.

Using the convenient transportation system of street cars and cable cars over the steep hills, she easily traveled the entire city. One of her favorite places was Golden Gate Park, a truly natural urban park. When the Panama-Pacific International Exposition opened in 1915, commemorating the completion of the Panama Canal, she roamed the grounds for hours. She was fascinated by the Greek architecture, the streams and fountains, and what had been accomplished in such a short time after the disastrous earthquake.

When she was not seeing Dr. Brown, they kept up a lively correspondence. He recognized her strength of character and encouraged her to find those depths and not rely on others. While their constant correspondence produced letters of encouragement, Brown did not want her to become totally dependent on him. His letters emphasized she should only see him when she had real needs. "I do not think seeing you as a patient regularly is fair to you or me."

Brown suggested she find outside interests, perhaps taking classes that might interest her. Two classes drew her attention -- painting and photography. Her art focused on the beauties of nature. While she enjoyed painting, she was more intrigued by photography. As with every endeavor she undertook, Mexia studied every facet of photography, taking her camera everywhere. She also continued to write, experimenting in many different genres from short stories, to plays, to nature articles. She recalled the lengthy letters she once wrote to her father -- and how she enjoyed putting pen to paper. Now she experimented with new approaches -- not so much to sell anything but to gain confidence in herself.

In 1917, she took a major step, spreading out beyond what she had ever done in the past. She joined the Sierra Club, attending meetings and going on outings. The outings took her to Yosemite National Park and other sites throughout the Sierra Nevada. The club's emphasis on nature and the environment originally attracted Mexia, and now she wanted to be part of this new movement. Through the club, she also began to make new friends. Slowly, Mexia gained more confidence in herself, finding she was more interested in her surroundings than ever before.

Mexia demonstrates her new self-confidence (and daring) in Yosemite National Park.
Photograph Courtesy of California Academy of Science, San Francisco

The Perfect Specimen

> ### The Sierra Club
>
> John Muir founded the Sierra Club May 28, 1892. Its original intent was to be a California association of men and women devoted to exploration, enjoyment, and preservation of the Sierra Nevada mountain range. Today, the organization's membership totals over 750,000 located throughout the country. Twenty-nine offices govern the business of building an environmental community, litigating when ecological interests are threatened, lobbying state and federal legislators, and educating the public at large. The expanded statement of purpose now reads: "To explore, enjoy and protect the wild places of the earth; to practice and promote the responsible use of the earth's ecosystems and resources, to educate and enlist humanity to protect and restore the quality of the natural and human environment, and to use all lawful means to carry out these objectives."

The benefits of joining the Sierra Club were numerous. Hiking in the outdoors provided a physical outlet for Mexia to strengthen her body. Association with other hikers also gave her new social outlets. Many of the hikers and club members became lifetime friends. And certainly the outings, whether they were day hikes or camping trips, served to trigger Mexia's interest in nature. Often, prominent professors came along to speak about forestry, bird, animal life, and history.

Mexia especially enjoyed the deep quiet of the redwoods. These amazing, gigantic trees had been on Earth for centuries. She could walk among the giants for hours, speculating on how long they had lived. In 1918 and early 1920, it was a long distance to the Northern California coastal area so few people traveled to view these redwoods, and interfere with the solitude.

Durlynn Anema, Ph.D.

Mexia first visited a redwood grove in Montgomery Woods State Preserve, Mendocino County, California. *Photo Courtesy of Evan Johnson, Save the Redwoods League.*

She first visited a grove of redwoods, now called the Montgomery Grove, in 1918. These magnificent redwoods were being threatened, both by the timber industry and by natural disaster, namely earthquakes. This visit motivated her to become a member of the newly formed Save-the-Redwoods League, gladly paying $2.00 annual dues. In a letter to the organization after visiting Montgomery Grove, she wrote, "I am heartily in sympathy with any effort to save these trees, and wish to inquire whether it is only the trees in Humboldt County which are under consideration (to be saved) or groups throughout the State." She also wanted to be sure these trees would be saved from the "fate of the axe and shake." (See Additional Chapter Sources for complete letter.)

Acting upon her advice, the League's first Secretary Newton B. Drury, requested that the Committee on Redwoods Investigation consider whether the grove was worthy of protection. By 1920, Drury reported to Mexia that "cutting in the heart of the grove had been stopped through the efforts of officials of the League and the citizens of Ukiah, California." (See Additional Chapter Sources for complete letter.)

> Permanent protection of these trees was not secured for another twenty-five years. With donations and purchases, the grove was enlarged to 1,142 acres and is now called Montgomery Woods State Reserve. It is one of the last remaining places in Mendocino County where the public can revisit the wonder of the ancient coast redwood forest as Mexia experienced during her first visit in 1918.

Mexia worked with officials of the League, suggesting fund-raising ideas for preservation and recruiting new members for what she considered a worthy organization. An article in *National Geographic* revealed the grandeur of these redwoods to the American public and helped raise funds for preservation. The objective of the Save-the-Redwoods League was to rescue areas of primeval redwood forests from destruction. They also wanted to cooperate with state and national park services to establish redwood parks. Between 1920 and 1928, the league purchased many redwood groves that formed the core of the California Redwood State Park System and later the Redwood National Park.

At this point, while Mexia realized all Reygadas had done for her, she knew she could never return to Mexico and her old life. She also was upset at his lack of financial acumen -- he was bankrupting the ranch. Theirs had never been a strong marriage, so with reluctance he agreed to a formal separation. They remained separated, but records are not clear as to whether they ever divorced. She sold her ranch and took back her name.

The years had taken their toll on Mexia. It was now 1920. She was almost fifty years old and facing an unknown future. Her only certainty was her need to learn.

Chapter Four

Introduction to Botany

Mexia so enjoyed her time on nature outings that she wanted to explore more, especially the plants and trees which always surrounded her during her Sierra Club trips. While she was intrigued with the geological features of the landscape it was the living, growing items which most fascinated her. What is the process of their growth? Why are there so many species of plants even within the plants' genre? What more could she learn about them?

The answer to her questions came quickly. She would attend college.

In 1921, Mexia enrolled at the University of California, Berkeley as a special student. She was fifty-one years old, an anomaly in those days. Only young people attended colleges and universities. Mexia didn't care. She felt education was a lifetime experience and was thrilled to return to learning. She had loved school when she was young and now realized she should have continued -- but that was in the past and this was now.

The first course she took was natural history. While Mexia enjoyed learning about all organisms -- animal, fungi, and plant -- she was most intrigued by the latter. She remembered as a young girl that whenever she roamed the outdoors it was plants that most fascinated her. She loved everything from the brightly colored flowers, to the cacti in the desert, and even was fascinated by weeds which seemed to spring up everywhere.

A university expedition led by Dr. E. D. Furlong, curator of paleontology at the university, introduced her to botanical collecting. This introduction to plant life captured her attention as no other project ever had and convinced her to immerse herself in the study of botany. Now Mexia had a focus for the remainder of her life: she wanted to be a botanical collector and explore known and unknown places on earth in search of plant life.

Immediately, she took more courses in botany, and in 1922 joined a botanical expedition from the university to Mexico. This trip exposed her to botanical collecting in the wild. While she made only a few collections, she was so excited about the process she could hardly wait to do more exploring.

> Botanical collecting is a precise and laborious job.
>
> If a plant is herbaceous, or containing little or no woody tissue, a botanist will unearth the entire specimen. Woody tissue samples, because of their size and weight, usually require a pruner to clip a piece off. In certain instances, plants must be dissected in order to properly identify them. The plants are then pressed and dried, sometimes in the field and sometimes in a lab. (Mexia did most of her pressing and drying in the field). Old newspapers can be used to separate the specimens until they are offered to museums and herbariums, where they hopefully will become part of a permanent collection.
>
> This all was a process Mexia wanted to learn and pursue; because she loved details, botanical collecting was a logical path for her to follow.

When she went to the California Academy of Sciences in San Francisco for additional study, Mexia met Alice Eastwood, an important American Botanist. Mexia had already heard of Eastwood because of her prominence in the Botanical field.

Eastwood provided critical specimens for professional botanists and also advised travelers on methods of plant collecting. She was a self-taught botanist, first studying plants in

The Perfect Specimen

Colorado's Rocky Mountains while working as a schoolteacher in Denver. Her botanical studies were so impressive she was hired by the California Academy of Sciences in 1891 to assist in the Academy's Herbarium. The next year Eastwood became joint curator of the herbarium. By 1894 she was procurator and also Head of the Department of Botany. She also founded and ran the California Botanical Club.

One of Eastwood's contributions to the Academy was saving the critical foundation of the Academy's collection from the earthquake of 1906. While the Academy was rebuilding, she studied herbaria in Europe and Great Britain including the British Museum and the Royal Botanic Gardens at Kew, and several U. S. regions, including the Gray Herbarium and the New York Botanical Garden. In 1912, she returned to the Academy as curator until her retirement in 1949. Between 1912 and her retirement in 1949, over 340,000 specimens were added to the herbarium.

Mexia felt it was a great privilege to work with Eastwood. She often accompanied Eastwood on field trips to the coastal ranges and the Sierra Nevada. During each journey, Mexia learned more about botanical collecting until she hardly could wait to begin on her own. She felt fortunate to know that whenever she had a botanical question, she could find the answer by asking Eastwood.

Mexia never was bothered by her age or the age of the young students who went on the field trips and expeditions with her. For her, they were all seeking knowledge together. She also continued to take as many classes as she could, not necessarily for credit or for a degree, but to learn more about botany and all its ramifications.

The story below illustrates her comfort with young people and shows how far she had come in enjoying people without fear. The remembrance happened to Mexia and a freshman named John Thomas Howell who later became curator at the California Academy of Sciences.

In Thomas' words: "...my first remembrance of her was a truly memorable occasion when she and I got lost together in the 'wilds' of Stanford University. ... it was September 30, 1923 on a field trip with the Calypso Club, the student botanical club at U. C. We had spent the day ... exploring the floristic riches of Jasper ridge which lies in the Coast Range foothills to the west of the university buildings. Since it was quite a hike from the ridge to the railroad station in Palo Alto, Herbert started us on our homeward trek early enough for a pleasant leisurely walk. Mrs. Mexia and I concluded it would be no fun to sit around the station waiting for the train so we decided to stroll along at a slower pace, enjoy the beautiful autumn hills and arrive more nearly at train time. Separated from the rest of the club, we made a wrong turn and lost our way in the maze of roads about the houses, barns, and sheds in the back part of the 'Stanford Farm.' It was dusk by the time we reached the station and long after our scheduled train had departed. Eventually, it may have been Oct. 1 when we arrived in Berkeley (by a late train to San Francisco, streetcar to Ferry Building, ferry across the Bay, interurban to Berkeley, streetcar to our homes). It was my initiation to the Calypso Club and Mrs. Mexia never let me forget it!"

The Perfect Specimen

Howell went on to comment that Mexia accompanied the Calypso Club on many of its excursions. She acted as "chaperone" when the club went on longer trips that necessitated camping out a night or two.

In 1924, Mexia, with her marriage over and having sold her business interests, cut her ties with Mexico and reestablished U. S. citizenship, which she had given up during her many years in Mexico and her two marriages to Mexican citizens.

Eager to continue her education, Mexia in 1925 took a course in flowering plants at the Hopkins Marine Station in Pacific Grove, California, on Monterey Bay. There she met Roxanna Stinchfield Ferris, Assistant Curator at Dudley Herbarium, Stanford University. Ferris and her husband were going on a collecting trip to Sinaloa, Mexico to collect specimens for the Herbarium. Ferris invited Mexia to come along.

Mexia was thrilled. Not only would she be returning to a country she knew well, but she would be given the duties of a true collector. She totally understood one of the reasons Ferris asked her was because of her connections in Mexico. She would be able to get assistance and privileges not always available to outsiders. This did not bother her because she would have the opportunity not only to learn more about collecting but to collect on her own.

In preparation, she contacted Eastwood with a proposal. She saw no reason why she couldn't duplicate the collection she made and send it to the Academy. She assured Eastwood she already had a "good deal of experience in the collecting and preparing of Herbarium specimens," and would forward the specimens directly to Eastwood from Mexico.

> When Mexia wrote to Eastwood she was honest about why she was doing this.
>
> "There is no reason that I should not duplicate the collection made by Mrs. Ferris, and I was wondering whether the Academy would be interested in such a collection. I have had a good deal of experience in the collecting and preparing of Herbarium specimens, and Dr. Abrams, with whom I am doing some botanical work here during the Summer Session of Stanford, suggested that I write to you."

Collecting was expensive. The equipment and materials for pressing and drying had to be bought if Mexia was to collect on her own. She wrote another letter to Eastwood in August asking only for expenses, assuring Eastwood that if this was arranged, the entire collection would be the Academy's. Eastwood agreed to pay on a per-specimen basis when Mexia returned.

> The letter Mexia wrote to Eastwood said in part:
>
> "I am willing to collect for expenses only, and if your budget does not make allowances until the first of the year, I think I could manage to finance myself temporarily. . . under this arrangement all that I collected would belong to you and be sent direct to you from the field as soon as prepared."

The trip proved to be a disappointment. Working with Roxanna Ferris was not productive. Shortly after they arrived in Mazatlan, Mexico, Mexia realized she wanted to collect in her own way and did not enjoy the company of the Ferrises. She had been independent for so long that it was difficult for her to follow the orders and plans of others. The Ferrises agreed she should leave the expedition. Now Mexia was on her own!

Chapter Five

Now On Her Own

While making the decision to leave the Ferris' and collect on her own gave Mexia a feeling of independence, she knew this decision also would be expensive. She needed her equipment, additional materials, and some money for expenses including hiring a guide.

She wrote friends in California, asking them to send her equipment. Again, she contacted Eastwood, asking for any help, equipment, or materials the academy could spare. Mexia knew she was taking a chance on going by herself but also knew this was the only way she would determine if she could do this on her own.

> Botany collection is an expensive occupation due to the equipment necessary for pressing and labeling the specimens found. A drier and a plant press must be carried along with materials for pressing the specimens. The prepared collector also brings a compass, pruner, trowel, camera, and any other necessities for survival. All this equipment is a burden. As Mexia wrote after her Mexican expedition, "Transporting of equipment and specimens is a real problem in Mexico, where packing by animals is the only method of getting around, and that often difficult and expensive."

> Collecting is also expensive because botanists tend to make several duplicate sets when they collect. They do this so they can distribute their collections to several different institutions and herbariums. While most of the collection goes to institutions in the United States, the collectors also want to include institutions in the country where they are collecting. Duplicating also means that collections, which are irreplaceable, are less likely to be completely damaged or destroyed. In addition, duplication ensures that specialists for a particular group of plants will receive and identify their collections, providing a second opinion for identification.

Mexia planned to venture into the hills and mountains above Mazatlan, then down the coast. While the expense of doing this by herself was an obstacle, she was independent enough to realize she no longer could go on expeditions with other people. She felt that going at her own pace into places of her choosing would be more productive.

In the 1920's, women did not venture into unknown regions of the world. A few had explored the wilds of South America -- Harriet Chalmers Adams, Annie Peck, and Mary Blair Niles in the early 1900's. None had spent months in the wild by themselves.

Mexia's wait in Mazatlan for her equipment proved productive. She met a botanist who was happy to advise her about the area around the city. He explained where no collecting had yet been done, and helped her move her supplies through customs.

When her equipment arrived, Mexia was ready to explore on her own. She first went along the coast north and south of Mazatlan, collecting coastal forms of every flower and plant she could find. While she realized many of these probably had already been discovered, she hoped a few might be new specimens. Then she went into the foothills east of Mazatlan, to an elevation of about eight hundred feet, for more obscure collecting.

The Perfect Specimen

Quickly, she realized she had to find a guide. While she knew Spanish and could get along with the populace, she did not know all the trails or the best ways to go. A native guide was the answer.

The native Mexia engaged was familiar with the trails of the Sierra Madre east of Mazatlan. They went on horseback, climbing higher into the mountains. The starkness of the landscape fascinated her. She also realized collecting could only be done at certain times of the year -- after the rains had come and gone.

Two days later, tragedy struck. Mexia saw a specimen far out on a ledge above a cliff. She was not even aware of her next move; all she cared about was the specimen. She moved ever closer to the ledge. Her determination got the best of her, and over the cliff she went. While the fall wasn't a great distance, it was severe enough that she broke some ribs and injured the hand upon which she landed.

Fortunately, she had her guide. He led her gently to her horse, then helped her into the saddle. In a great deal of pain, she rode back to Mazatlan, where she learned she needed surgery. She had to return to California.

Mexia had obtained nearly five hundred species at the three elevations where she explored, but had hoped for a thousand more. One, the *Mimosa mexiae*, became the first of a number of plant species that were named after her. She had also learned that while she might be fifty-five years old, she had the endurance to survive the collector's rough, lonely, and adventurous life.

Reluctantly, Mexia left the western coast of Mexico and sailed home. As she watched the coastline of Mexico disappear, she vowed that as soon as she was well she would return.

Botany is an ancient science.

A. G. Morton commented that "Botany is indeed embedded in human history from its origins in two of the most ancient sciences of all, magic and medicine, through its long association with pharmacology, agriculture and horticulture, to its part in the exploration of the world, and in ensuring the supply of food and raw materials for the rise and maintenance of modern industrial society.

What began with the earliest human beings as a search for food led to the cultivation of certain plants in the form of crops, then advanced levels of agriculture, and finally maintenance of livestock. As the use of plants expanded, medicinal benefits of certain species were discovered largely through trial and error. Plants have since taken on additional significance as beauty and art, as anyone who has seen an orchid or a tiger lily knows...

...The diversity of plant life is extraordinary. Their shapes, colors, and sizes vary from minute to immense. The scientific division of plants starts at the class level, where those plants with flowers are separated from those without. They can be broken down ten times further according to a variety of botanical characteristics that are used to bring order to the mass of plant life.

Approximately 325,000 kinds of green plants alone have been described. New ones are constantly added to this list. The main reason plants have been classified is to divide those plants that are beneficial to humans from those that are not.

Botanical collectors are a vital part of plant life study. They not only confirm present species and genera (a closely related species), but their discoveries might uncover a new plant and/or use for that plant. Serious collectors maintain a rigorous pursuit that takes the practitioner into the wildest of environments.

CHAPTER SIX

HER FIRST LONG SOLO EXPEDITION

Mexia decided if she was going to Mexico she should do it right and not rush. Therefore, she planned to take at least six months in her endeavors, going not only to Mazatlan and the mountains surrounding that area, but possibly to Puerto Vallarta and the mountains outside that beachside town. She had heard very little botanical collecting had been done in the Puerto Vallarta area so she was curious about what her prospects might be.

However, her first endeavor would be in Mazatlan and the Sierra Madre mountains east of there.

Horses, dugout canoes, small boats, and foot power were paramount on Mexia's first major collection effort by herself. Although she was collecting under the auspices of the University of California Department of Botany, she needed additional funds. As Mexia wrote after her Mexican expedition, "Transporting of equipment and specimens is a real problem in Mexico, where packing by animals is the only method of getting around, and that often difficult and expensive."

She was able to obtain funds or promises of payment -- at twenty cents per specimen -- from several herbariums and natural history institutes, ranging from Harvard University to the Royal Botanical Garden. In late August 1926, she left San Francisco on a Pacific Mail steamer.

Durlynn Anema, Ph.D.

Four days later, the steamer reached the tip of Baja California and sailed east to Mazatlan, the port city of the Mexican state of Sinaloa. From the ship's railing Mexia could see the fertile inclines of the Sierra Madre.

At the pier, she was met by J. Gonzales Ortega, a civil engineer, amateur botanist, and good friend. She said "this gentleman knows the west coast of Mexico as few know it, and has been most helpful in advising me as to the best localities for collecting."

Immediately, Mexia set out by train for Tepic, Nayarit. This quaint old city was in a fertile valley of the Sierra Madre mountains at an altitude of 3,000 feet, which was an advantageous altitude for collecting because it was out of the coast's humid heat. Despite being on the new Southern Pacific railroad line, Tepic was largely unspoiled by civilization, much to Mexia's satisfaction: "The streets, when not consisting of mud holes, are cobbled and bumpy, with houses and walls of adobe on either side, but the houses are far apart, each set in its garden, or huerta, and the red tiled roofs nestled picturesquely among the tender greenery of the bananas, or the dark glossy green of the omnipresent coffee plants. Even the adobe walls become things of beauty in this ideal climate, for they are covered with a garment of Maiden-hair fern, while above them droop the branches of flowering or fruiting trees."

Ortega had provided Mexia with a letter of introduction to an official in Tepic. With this help, she was able to hire a mozo (guide) named Mauro and two horses. Her fluency in Spanish was a great advantage because she could communicate with everyone she met. Each morning she and Mauro rode out from the town to explore and collect. She was amazed at the abundance of vegetation, so much "it was hard to know where to begin to collect, and still harder to know when to stop."

Each day, Mexia and Mauro would start on a road, then cut off onto a small trail which led to the higher elevations. Mauro carried the field-press. Mexia got off her horse to collect and pack away all the plants she could find within a specified area. Then she mounted her horse and rode to another location.

The Perfect Specimen

Along the way they stopped to eat the wild guavas. Mexia was amazed at the abundant ferns growing along the trail and in every ravine: "Convolulaceae of every size and color were everywhere along the hedgerows and clambering over shrubbery and small trees, very ornamental but a terrible pest to agriculturists...The wild fig-tree, *Ficus mexicana Miqu*, here grows to huge proportions. As the green fruit hung high, Mauro deftly lassoed some fruiting branches for me."

Usually by three in the afternoon she had collected all the specimens she could handle. Immediately, she wanted to get them into the presses to dry them. Mauro learned how to help her, which she appreciated. Even so, the task often was not completed until after nine o'clock at night.

On the trail to the village of Jalcocotan, Mexia enthused over specimens that grew wild in Mexico while gardeners in the United States had to nourish them as flowers if they were to thrive. She found cosmos and scarlet dahlias everywhere. Zinnia, verbesina, and hibiscus grew wild in the denser wood.

Finding a dense, overgrown trail near Cerro de San Juan, a spur of the Cordillera, Mexia talked Mauro into taking it. He informed her it had not been used since the mountain had been a stronghold for bandits during the revolution. This only made the route more alluring to Mexia.

As they traversed the narrow path, she delighted in the strange and fragrant deciduous trees and the ferns and mosses growing in the dense shade beneath. On their descent the trail was more level and open, leading her to find a new species of shrub (called pie de pajaro by Mauro and officially named *Deppea macrocarpa*).

Mexia left Tepic after two weeks of collecting and went southwest by train to Ixtlan del Rio, which was higher into the mountains. She hired a new mozo, Juan, whom she felt was very competent. They rode up the sides of the mountains until they reached the lower fringe of the oak and pine belt. Of the openings of the wood, she wrote, "Many composites were now in flower. On the more sunny slopes we found cacti, among others, a red-fruited Cereus (species). This latter strikingly demonstrated in what manner the slabs lose their spines and shape and become trunk-like as the cactus approaches tree form in age."

Mexia was accustomed to the food of Mexico because of her years of living in that country so knew she would have frijoles and tortillas at every meal. She also appreciated the wild game her mozos often shot and fixed over the open camp fire, but was less enthusiastic about the dried shrimp villagers snacked on like peanuts.

After returning to Mazatlan to deposit her collection, Mexia took a short train trip to Los Labrados and the jungle around Marisma for four days. Returning to Mazatlan, she then went south via train to the village of Ruiz.

Looking around the area surrounding the town, the mountains to the north looked inviting, so she decided to explore them. However, the villagers told her to take an escolta (guard) if she wanted to go. Mexia didn't like this idea: "I privately think an 'escolta' ... could be a great nuisance standing around looking at me while I collected and put the plants into my press. I would feel as though I were under guard, but I am terribly intrigued with the idea of the Indians and the mountains. They are not savage Indians like the Yaquis, just shy and afraid of strangers." Because she could not hire a guide without the escolta, she realized she had to forego the trip.

To date, collecting specimens seemed a limitless task to Mexia. She felt at times neither her materials nor her two hands could "keep up with it. My driers get all filled up and still numberless plants sit and look at me and announce that they are still waiting to be collected. I work to my capacity with what green help I can find in each new locality, and then everyone tells me of all sorts of delightful points to which I have not get gone, and of all kinds of plants I have not yet come across! It is terribly trying to a greedy collector like myself."

The birds also entranced her. She wished she could have an ornitological companion to identify the multitude of species surrounding her.

THE PERFECT SPECIMEN

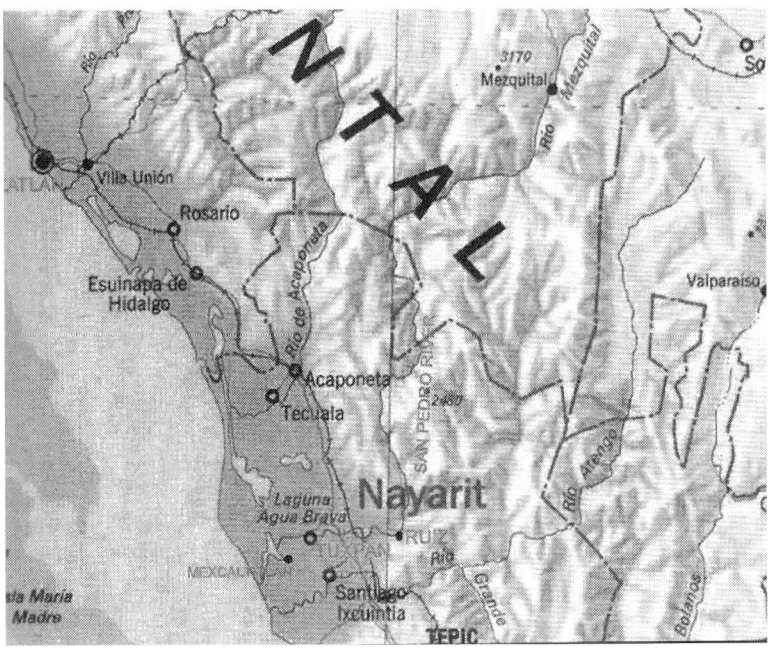

The trail Mexia took down the San Perdo River and through the Sierra Madre to the costal plains.

A dugout canoe took Mexia on her next exploration down the Rio San Pedro from Ruiz to Tuxpan, still in the state of Nayarit. This river is one of the longest rivers in Mexico, starting in the mountains of Durango and flowing through part of that state then into Nayarit before draining into the Pacific Ocean. This river flooded regularly so the surrounding soil was highly fertile alluvial deposits. "Crops (and weeds) grow almost overnight," Mexia later wrote.

From Tuxpan, another dugout canoe took her to Mexcaltilan. This village of shrimpers sat on an island in the lagoon formed by the delta of the river. Mexia had decided this was a logical place to find new specimens with little difficulty. Dugouts and canoes were the only means of transportation. The river, although broad and swift, was shallow in many places. The canoes were hollowed out of a single tree by hand in the mountains, then brought to the lower lands for sale. In the rainy season, all freight and passengers were carried by these canoes.

Mexia's new mozo, Antonio, piled all her driers and equipment in the middle of the dugout. Then she sat in the prow and he stood astern with a long pole and a broad paddle for the deeper places. Beautiful birds lined the sand banks -- black-necked stilts, great blue herons, sandpipers, and even turkey buzzards to clean up the dead animals, birds, and fish. However, she hardly had time to view the birds between collecting specimens and fighting mosquitos.

The mosquitos had become fierce: "The mosquitos and 'jejenes' (biting gnats) arise from the water in clouds! I suppose they will have an extra good time eating up the poor 'Gringa' ... I am instructed to take quinine nightly as a precaution against the deadly mosquito. So far I have had no accidents, illnesses, nor anything except bugs, and everywhere I have gone I have been received with courtesy and helpfulness."

Mosquitos were not just an annoyance but also had the potential to transmit disease, especially malaria. This was not the only health risk Mexia took. She had been forewarned of an outbreak of smallpox in Ruiz, so she was inoculated by a doctor in Mazatlan before she left. However, the inoculation did not take. She was not afraid, even though she had never had smallpox -- and she never did get it.

As Mexia approached the village, the setting enthralled her. "The village was on a small island in the vast shallow lagoons which stretch along this coast for leagues and leagues. Lovely blue water-lilies ... grew by the acre and the Corpus Christi ... lifted its stately white flowers and spread its immense pads in the sunshine. The lagoons, fresh to brackish, were everywhere broken by what looked like wooden islets, which when approached turned out to be association of water-loving trees: Mangrove ... the Buttonwood ... Phyllanthus acidus, Skeels and others, all growing in the shallow water. In places these trees were so smothered by vines ... that they lost all tree shape and became but living green mounds."

The Perfect Specimen

Her review of the village of Mexcaltitlan was not so glowing. "It was the most miserable little shrimp-fishing village imaginable and I now quite understand why a despicable sort of a person is called a 'shrimp.' You know the types: pigs, children, dogs, mosquitos, etc. all mixed up together and surrounded by an aura of decaying shrimp. My heart sank away down when I surveyed (and sniffed) the village and I began to wonder why I ever wanted to collect in far-away spots anyhow; but there I was and I could only make the best of it. I had a letter to the 'big man' among the village inhabitants and was received with the usual courtesy of the Mexican, be he rich or poor. Their best is always at the disposal of the stranger, yet their best occasionally seems to us...pathetically little."

Her accommodation was a back courtyard. She had been offered one side of the room where the family slept, but preferred some privacy. They placed her cot in the yard as far away from a mother pig and ten piglets as possible. However, her cot ended up being beside a hollow where water was thrown out, and each night the pigs grunted and splashed in the water.

A canoe took her out the next morning to collect. Mexia quickly collected all she could from the mangroves, which are water-loving trees. She was fascinated by the landscape. "The lagoons stretch for leagues and leagues and leagues up and down this coast and everywhere the surface of the earth is covered by clear, warm shallow water. In this area are beautiful, densely wooded islands of dark, glossy-leaved trees, which, when approached are found to be rooted below the lagoon surface, so there is no real land. Narrow, intricate channels wind between these tree clumps and occasionally they fall away on either side forming miniature bays and inlets ... the whole is like a beautiful carpet, for the great pads are dull green and carmine, and they form intricate designs over the clear, smooth water, everywhere starred by blossoms."

Mexia found not only collectible plants but a lustrous red-bronze bird called a jacana. She was thrilled but noticed her canoeist wasn't impressed. He could only tell her that these birds were very good to eat.

She remained at the village for two days but found little of value. Her return in a canoe was more difficult than the trip down to the village. It rained steadily and hard. Mexia crawled under the canvas covering her specimens and driers, and endured the snaillike pace.

It became extremely dark, and poling upriver against the current meant progress was by inches. They went aground constantly, at which point the mozo had to get out and shove the dugout into deeper water: "It was eerie enough, the treacherous, unseen river, the heavy, low-hanging clouds, the banks only guessed at by the fireflies' fitful sparks, the occasional glint on the stream of the light from nowhere that but makes the water blacker, the 'plop' of heavy, jumping fish, the long-drawn howling of far-off coyotes, and the unreliable individual with whom I had to deal."

After a weeklong stay in Tuxpan, Mexia was ready to leave the mountains of Mazatlan. Perhaps more valuable than the thousands of specimens she'd already collected was the experience gained surviving in the wild and running her own expedition. Now she would find a steamer going to Puerto Vallarta to see if she actually would be one of the first collectors in that region.

CHAPTER SEVEN

INTO THE SIERRA MADRE

When Ynes heard that few collectors had ventured to the Puerto Vallarta area and the Sierra Madre mountains east she was determined to go there before she returned home. However, getting there from Matzatlan was harder than she realized.

She discovered that Puerto Vallarta was such an "out-of-the-way" place that few steamers went there. After returning from her collecting trip east of Matzatlan, she had to wait five days for a steamer, which was overdue, to take her south to the small town. She also understood that leaving Puerto Vallarta might be difficult because steamers went there so seldom, so she probably would have to return to Mazatlan via the mountains, a four to five day journey by mule.

When the steamer arrived at Mazatlan there was cabin space for only sixteen passengers -- yet fifty-five people were on board. She stayed on deck, sleeping on cots provided. She said later said she "slept more comfortably than in a stuffy cabin as the night was warm and pleasant with every tropical star twinkling."

The trip took a night and a half day down the coast before entering Banio de Banderas (Banderas Bay), then hours more to cross this bay on the dividing line between the states of Nayarit and Jalisco. Her first impression was of "white toy houses clustered on a narrow ribbon of sand just in the shadow of the steep-pitching heavily wooded mountains. ... I alone was to be

Durlynn Anema, Ph.D.

abandoned in this most lonely spot where all sign of human habitation was dwarfed by the vast panorama of sea and sky and mountain."

The Sierra Madre frame the huge central plateau area of Mexico. Made up of three parts -- the Oriental, the Occidental and the del Sur -- the mountains climb from 6,000 to 12,000 feet, set steeply and covered in thick vegetation. Traversing them is no simple matter.

Dugout canoes greeted the ship upon arrival. Only Mexia disembarked. She took a deep breath as her equipment was unloaded into the canoes, more afraid for it than for herself. When they reached the beach she had to chuckle because of all the "cargadors" trying to help her. Fortunately, because she spoke Spanish, she was able to assemble a crew to carry her equipment.

Puerto Vallarta's Banderas Bay where Mexia began her trek into the Sierra Madre. *Photo Courtesy Durlynn Anema.*

The Perfect Specimen

She stayed in the house of a merchant, Mr. Guereno, who had invited her through mutual friends. The house was two stories with a wide corridor around the patio. The patio was enclosed on three sides with the fourth open to the sea breezes and view of the bay. While the weather is hot during the summer, in December it was pleasant for which she was grateful.

Mexia quickly discovered that the rumors about this unsampled terrain were true. The wilderness in this secluded, unspoiled area was immediate, and Mexia only had to go to the end of a small street to collect. Mr. Guereno already had found a mozo for her -- Trinidad.

Her collecting habits amused the villagers. They could not understand why this woman, who seemed to have money, climbed the rough mountains. All this, she wrote to friends, was "indubitable evidence (to the villagers) that I must be somewhat touched in the head, but then (they) never know what crazy foreigners will do anyways."

One day Mexia climbed high above the village on a mountain called Cerro de la Cruz, which had a large cross at the top. Her most exciting find was a tree with huge leaves in rosettes.

It was called the hincha huevos (or egg burster) by the natives, who believed an egg coming in contact with any part of the tree would burst.

Among the plants she found was lantana, which was abundant and growing wild. She laughed at this discovery because gardeners in the United States pampered it in their gardens. This was yet another example of how different the environments of the U. S. and Mexico truly were -- not only in the horticulture but also in the people and their society.

Climbing around the precipitous hills rising abruptly from the river's edge, she found a small tree later named for her -- *Eugenia mexiae Standl*. No matter where she looked, discoveries abounded, until she hardly had enough room to pack all the specimens down the mountains.

When Mexia reached the top of Cerro de la Cruz, she saw the Ameca River winding "like a silver ribbon through an alluvial plain forming a sort of delta which it had built far out into the bay." She was determined to make a trip to this region. But first she wanted to go even further into the mountains surrounding Puerto Vallarta.

To do this, she hired Pedro, a woodcutter who lived in the mountains. They used burros to reach his house. Ynes description of their journey: "....my driers were packed on one burro, my cot and stuff on another and I perched on a third... Now riding a donkey on an "aperejo" as the straw-stuffed pack saddle is called is not all it is cracked up to be. The saddles are hard and rough and there is nothing to hold on to. It was not so bad on the level beach trail but when we began to climb I kept wanting to slide over the donkey's tail until finally I decided that walking would be easier than trying to hang on to that thing, and off I went."

When they arrived at Pedro's house, far into the mountains, she was greeted by his wife and nine children. Although they wanted her to stay in their small hut, she insisted she needed to be outdoors. She did this for two reasons: she did not want to impose on their hospitality, and she needed time by herself, something she treasured throughout her life. She camped in a banana grove, which she said was certainly different from the pine trees she was used to in California.

Her location was Cruz de Vallarta, at an altitude of about 2,300 feet. She stayed five days, collecting with the assistance of two of Pedro's boys. The volcanic mountains were a strenuous climb for her, later writing that many of them seemed to be "standing on edge." She was happy to have the boys along as they eagerly climbed trees she felt she could not climb. They picked all the leaves she wanted although they did not understand why she wanted leaves.

Before leaving Pedro's family, she took photos of them. While she paid them for her stay, she also realized the photos were one of the greatest gifts she could give anyone in this region, because they so seldom saw a camera. Her only regret was she could not take more photos, but she did not have much film to spare.

THE PERFECT SPECIMEN

Mexia returned to Puerto Vallarta just before Christmas and was thrilled to find a packet of letters from home. It was the nicest present she could receive.

> Christmas dinner was spent with the Guerenos. There she learned distressing news about her first mozo Trinidad.
> When they were collecting she had asked him to get some specimens from the "Hincha Huevos" tree. This collecting had affected him greatly. He had a high fever, his face was swollen beyond recognition and his hands were useless. The tree was a Comocladia, a relative of the U. S. poison oak. As she wrote: "It had poisoned poor Trinidad while it had not affected me at all."

At Christmas Mexia also learned the return steamer would not be touching at Puerto Vallarta for some time. This meant she would have to pack out through the mountains. She decided to collect as high in the Sierra Madre as she could before thinking about her return. As her base she chose San Sebastian, the highest town in the mountains.

On the road to San Sebastian. *Photo Courtesy of Durlynn Anema*

Durlynn Anema, Ph.D.

Los Robles, the town before San Sebastian. *Photo Courtesy of Durlynn Anema*

San Sebastian was an old silver mining town nestled in a valley of the same name that rested 7,000 feet above sea level under La Bufa, the highest peak of the region. This valley was just below the frost line, so on the overhanging crests the temperature often dropped below freezing. The mines were no longer profitable to work, so the population had decreased with only a few of the old families remaining.

She and her packers camped out during the trip, so it took three days to reach San Sebastian. Mexia often would go to the rear of the pack train taking her up the trail to view the beauty around her. What she especially enjoyed were the butterflies. "These are enormous pure white ones, and these do not seem to have the rather swift, erratic flight of most butterflies, but float along in a slow-swaying motion as if they were but bits of snow-white tissue-paper blown on the errant breeze. Another kind have long, rather narrow, pointed wings, and are striped zebra fashion from tip to tip of wing in yellow and black. These fly low, and flit from flower to flower, never pausing long in one place. Then there are little blue ones and yellow ones, and gorgeous big fellows in browns with fantastic patterns of many colors on their broad wings."

Gazing at the steep, jagged mountains, Mexia knew she would have some rugged hiking ahead of her. Many narrow deep canyons made by the streams tumbling down from the crests, cut through the mountains. The canyons were crowded with deciduous trees and shrubs, while the slopes and crests were clothed with pine and oak forests.

"The varieties of oak are legion, among them some of the largest and most stately oaks it has been my fortune to see. The pines are also of many species, all that I found but one, being five-needled, and quite different in habit from those of more northern climes. They have not the pyramidal form so marked in our conifers but are quite umbrella-shaped with spreading branches and mostly open foliage, the very slender flexible needles fully fifteen inches long. The tufts of needles ripple in the breeze and catch the glint of the sunshine as our shorter-leaved pines cannot do."

San Sebastian Plaza. *Photo Courtesy of Durlynn Anema*

Upon arriving in San Sebastian she was greeted by a mother and her two unmarried daughters who lived in a rambling old house built around a patio full of flowers and fruit trees, and an orchard with coffee trees and more fruit trees behind it. She had quite a time persuading them that she wanted her cot in the wide corridor of the patio. The family thought it was most inhospitable of them to allow her to sleep outdoors. Not only did she enjoy her sleeping accommodations but the luxury of all the fruit she could eat simply by picking it.

The houses in the village were built of adobe with tile roofs. She gloried in the birds, especially how they found the hollow roof tiles and uncovered beams for nests. She commented that "while the human population had dwindled the birds show no sign of doing so."

Ynes found San Sebastian a delight. "Just behind the house rises a steep hillcrest, pineclad to the summit with the trees silhouetted against the bluest of skies and gilded with the morning sunshine. The sky is so clear here and the slope so close that I can make out every bunch of pine needles, as well as the flowering shrubs beneath them. Surely 'my lines are cast in pleasant places,' even if I do get a bit homesick now and then."

From San Sebastian, Mexia climbed higher to the remains of an ancient mining village, El Real Alto.

"Talk about the primitive!" she wrote from the village. "I am beyond electric lights, beyond kerosene lamps, beyond candles, beyond even dips (sic), and writing by the light of a fat pine torch.! And the funny part of it is that it gives a very good light, so our ancestors were not so badly off, after all. Better than that, it gives off appreciable heat also, which, as I am on the roof of this part of the world, is a very welcome adjunct."

El Real Alto was bright and warm in the sunshine during the day. But when the sun dropped into the west, the chill crept in and the nights were bitterly cold. The Indians put out water in little wooden troughs during the day. At night it froze solid, at which point they carried it down the mountainside to San Sebastian to make ice cream.

The Perfect Specimen

> El Real Alto was built by the Spaniards in the 1500's because of the silver mines in the surrounding countryside. The mines were in continual production until the early 1900's, when the decrease in the price of silver made them unprofitable. The people remaining there hoped the mines once again would open, which had not occurred at this point. The old mission church was one of the earliest in the region and plainer than later ones. It had a painting of the Trinity from the early seventeen-century that had been brought from Spain.

She stayed with a widow with three children -- a poverty stricken family. Yet, Mexia wrote, "it is as clean and tidy as can be. There are no unwelcome six-legged inhabitants, and Guadalupe makes me the most tasty of little white tortillas and all the simple dishes she can."

The first day there Mexia and her mozo, Jose, climbed Bufa. The climb was not easy, but she was unafraid, even at "two or three ticklish places." When they reached the peak, her effort was rewarded. The view included the mountains around them, the valley of San Sebastian, and the town below, which looked toy sized. Beyond that was the Pacific Ocean. Following custom, Mexia inscribed her name on the book of ascents, a record of all who had reached the peak. In this case her record was engraved on the broad leaves of a live maguey, or agave plant, which grows in very arid conditions, much like a cactus. The maguey clings to rocks and is used in over one hundred products.

A steep slope towards the south -- so steep that no cattle had been there -- proved to be thick with interesting vegetation. The oak trees were thickly covered with orchids: "Some very prickly, thistle-like plants filled my fingers with little spines. On the peak, a milkweed grew highest of all the herbs, and a number of the mints were in flower in lilac and blue and scarlet."

Collecting and drying became so fascinating Mexia and Jose lost track of time (as usual). Jose kept saying they never would find their way in the dark because of the faint trail.

Mexia later said, "I did not know I could go down-hill so fast!"

They reached El Real Alto after dark and were greeted by a relieved Guadalupe, who thought they were lost. Guadalupe's concern impressed Mexia. She commented several times during her Mexican travels that everyone she met was so cheerful and hospitable in spite of their stark poverty.

Mexia had lived in Mexico for years but never had traveled into the mountains. While she had appropriate clothing and bedding, she noted that the family shivered under a thin cotton tablecloth at night, all huddled together. Many families could not even buy frijoles, except for a treat, and had to eat tortillas for every meal, with no meat at all. She tried to help them as much as she could, amazed at how they wanted to share with her when they had nothing.

After spending five days in El Real Alto, Mexia and Jose returned to San Sebastian. She sorted and labeled her specimens, and readied them for sending home. However, she missed El Real Alto and decided to return for another collecting round.

On February 27, Mexia and Jose returned to the tiny village. This time it was much colder. She wished she had brought more warm clothing, especially at night. Each morning, she welcomed the warmth of the first sunlight and readied for a day of collecting: "It is a great wilderness to explore, the tiny trail at our feet the only visible reminder of man, and it is so friendly and inviting, not stark and forbidding and awesome as our more northern mountains often are and I long to wander over it until I am intimate with each deep gashed canyon and each tree-clad upflung peak."

The first few days they stayed close to the village area. The stream beds were rocky and dry. Mexia wished they could find a stream with real water in it, not just rocky outcroppings of former streams. The villagers told her about such a stream, El

The Perfect Specimen

Jaguey, that was only two to three hours away. They started off early in the morning along the indicated trail to find the stream. This was an area Jose was not familiar with but persevered.

By eleven in the morning, Mexia realized they needed water for lunch. They had started on a gradual decline that wound around and around as it descended. Twelve o'clock passed. Three more hours of wandering brought them to the brink of a canyon. They looked into a deep gorge in whose bottom they could hear, if not see, running water. They plunged right down the gorge on a steep zigzag trail. At three-thirty, they reached a cold, clear stream fringed with alders and beautiful willows. Mexia later wrote: "We did not know where we were, but we were definitely somewhere."

Mexia had not been able to completely bathe while in the village, so she took advantage of this opportunity. "It was a 'dip' all right," she later wrote, "for the water was icy and the sun westering, so you can imagine that my scrub, while thorough was hasty to say the least."

After lunch, she began to collect. They had come too far for her to leave without specimens. Mexia found the first alders she had seen in Mexico and some herbaceous plants. When Jose insisted they go, she hated to leave, but knew they must. At five o'clock, they started up the slope.

"It grew black night before we were half way home, of course, and many an uphill mile still to go...It was remarkable to me how much better Jose could get along in the dark than I could. I can easily keep up with him in the daytime, but he soon got so far ahead of me that he had to sit down and wait a long time at a herdsman's shack. Then I rather intimated that I preferred he would keep at least within hearing distance."

They continued walking because they couldn't stop due to all the rocks. Then Mexia took a stumble in a rocky gully, hitting her shin and tearing her ragged clothes even more than they already were. Right after that, Jose fell head first but fortunately was unhurt.

Finally, Jose agreed that they must have a light -- the night was pitch dark. With his machete, he made two fat pine torches, which made the journey easier. The only problem was that the torches threw out showers of sparks, and in the dry forest that meant they had to keep stopping to put the sparks out, which was time-consuming.

As they saw each familiar landmark, they knew they were closer to home. Guadalupe had been close to sending a search party for them when they arrived.

The next day Jose and Mexia discovered they had taken the wrong trail at the abandoned shack, going five miles out of their way. A few days later, she decided to try again. They went back to the Jaguey, both because it was a beautiful stream, and because she was certain the area would yield many more plants (as well as thrills). She noted they had found "some of the most magnificent ash trees along this stream that I have ever seen and alders and willows and other trees that we commonly think of as growing only much further north, and of course I could not leave these uncollected."

As they went along the rough terrain they found some ferns whose "fronds were four to five feet long and the roots large and heavy in proportion." She knew the load would be too heavy for Jose and said later: "It was like losing teeth to leave those ferns, I know I will never have a chance to collect them again, but I took only one . .. of the big ones, and the smaller ones and reluctantly tore myself away (I still mourn their loss)."

As usual, the day passed quickly and light became precious as the terrain grew more difficult. At the end of the day, they came to a stream plunging twenty-five feet over a waterfall. Jose decided to work his way up the north wall, clinging to the crevices. Mexia followed him. Then he disappeared.

As she worked her way upwards the walls broke away in rotten rock and loose shale. Every time Mexia found occasional tussocks of coarse grass to cling to, she rejoiced, especially when she heard the loosened rocks crashing into the gorge. Jose kept yelling at her but she never turned around, just kept climbing upwards.

Memories of her previous fall kept her focused on her upward climb. Then she saw a tree jutting out a bit. She pulled herself up on the root and "just hugged the trunk as hard as I could. It was

The Perfect Specimen

firm and solid, it neither slipped nor gave and never have I embraced a tree trunk with more fervor!"

Mexia scrambled to the next tree and sat down to breathe. She was high up on the slope but on the opposite side of the canyon from Jose. He was hopping up and down and shouting. She could hardly hear him but decided he wanted her to come down the canyon again and up his side. She later wrote, "This I firmly and irrevocably refused to do. I went up that cliff because I got started and could not stop but nothing but the direct necessity would make me go down it."

Jose kept insisting, saying that if she stayed where she was and tried to go on, she would get lost. Mexia agreed with him. Finally, he said he would come to her. She was very appreciative.

Mexia anticipated him picking out an easier route than she had taken. With that in mind, she sat down to wait for him (one arm still around a pine tree). However, when he finally arrived, blowing and puffing, he said he had followed her tracks. Then he said, "I do not see how the Senora made it." He had slipped twice and thought he was gone (he also was carrying the heavy plant press). He dug toe and finger holds with the botanical pick he was carrying and continued to climb until he made the grade.

It was pitch dark but that did not inhibit Jose's sense of direction. Finally, he found a cattle path and they walked the three miles back to El Real Alto.

Jose was pleased when Mexia said they would return to San Sebastian. He also told her how impressed he was at her walking ability. Members of the upper classes did not walk in Mexico, only the poor. While she wasn't a fast walker, she had a great deal of endurance. He went around boasting of her prowess: "Another like the Senora I have never seen," he said.

Ynes had come to love San Sebastian and hated to leave.

She later wrote about this charming town.

"No wheel has ever desecrated its well-cobbled, winding lanes and the mountain passes leading to it bear only 'horseshoe' trails as they are called. One hears but the click of hooves or soft pad of sandals coming and going. The little plaza is flower planted and open to the sun and breeze, and the two 'policias,' nominal guardians of the peace, seem to have as their only task the watering and sweeping of its little winding paths.

"The church is large, and as in all Latin-American towns the point to which all streets converge. It is not so old by far as the old chapel at Real Alto and more massive than beautiful. It is kept open, but by order of the Government no services are held there at present.

"The remnant of the old families still live in the picturesque red-tiled, finely built old houses, ancient Spanish style with iron-grilled windows toward the streets and whose only entrance is through the great 'zaguan' heavily doored and barred."

THE PERFECT SPECIMEN

It was three days by pack train to the nearest railroad station. Her new friends wanted her to stay longer but she knew it was past time to go.

With regret, Mexia realized she had to leave San Sebastian with her 33,000 specimens and head for San Francisco. She had grown very fond of the town and its quaintness -- but knew she had to take her specimens home for distribution. She was, as always, determined to finish what she had set out to do.

> Author's Note:
>
> In 2006, I visited San Sebastian as part of a day tour. When we went into the museum, an older woman in a wheelchair told us about the artifacts and town. After she finished, I asked our guide (the woman spoke only Spanish) if the woman knew anything about Ynes Mexia. How thrilled she was to say yes. Although she was quite young when Mexia was in San Sebastian, the woman remembered her and how all the people of the town loved her. So not only did Mexia love this quaint, lovely village but they loved her.

Chapter Eight

Exploring Alaska, The Last Frontier

Mexia returned from Mexico with nearly 6,600 different plants -- lichens, mosses, ferns, grasses, herbs, shrubs, and trees. Included in these was one new genus and about fifty new species, ten of which were determined by the Field Museum of Natural History and six by the Harvard University Botanical Museum. Her reputation as a collector was growing.

On her return from Mexico, she was paid the standard rate of twenty cents per specimen for what she'd brought back, plus additional funds for photographs. Her quality work prompted the British Museum of Natural History to fund a trip to Alaska. Dr. Francis W. Pennell, curator of the Academy of Natural Sciences in Philadelphia, promised that, based on the Mexico collection, the Academy too would be interested in her yield from Alaska. With these assurances, she was ready to head north.

Accompanying Mexia was Frances Payne, a young botanist paying her own way. She thought it would be exciting to collect with an experienced scientist and learn new techniques. Mexia was honored to be considered so important, and proud to become a mentor to the young woman, much the way Alice Eastwood had mentored her.

The women left San Francisco on June 9, 1928. They landed in Anchorage, then continued overland to Mt. McKinley Park (now called Denali National Park). This park had been established in 1917. Mexia knew that no botanist had fully explored the park, giving her an advantage. The summer season is short in this area, making it critical for Mexia to move quickly. She would have only until mid-September before the weather turned.

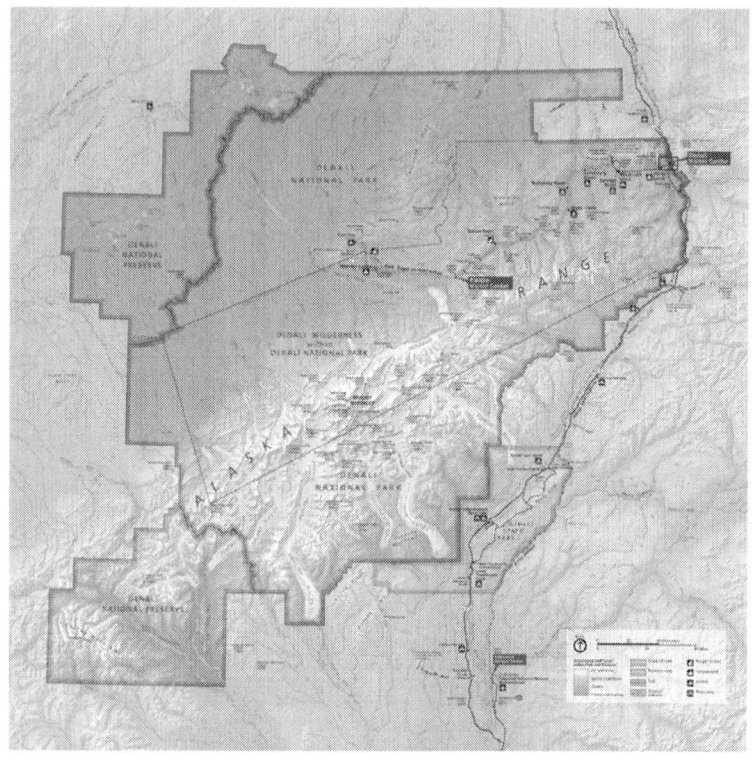

Map of Denali National Park (McKinley National Park when Mexia visited.) Her explorations spanned a 100 mile radius north and west from Wonder Lake with no roads west from the park headquarters.

The Perfect Specimen

Mt. McKinley at 20,320 feet, and the highest peak in North America, is the park's primary feature. However, there also are over six million acres of additional wildlands to explore. Grizzly bears, wolves, Dall sheep, and moose make their home here. Even before trekking into the wilderness, Mexia was certain she had made the right choice for collecting. (In 1976, the park was designated an international biosphere -- a laboratory for natural science research.)

Frances Payne, on the other hand, underestimated Mexia's adventurous spirit. She was not prepared for the rugged terrain and her mentor's willingness to tramp through it in search of the unfound. She quickly decided to stay closer to comfort, near park headquarters. Mexia, no longer so interested in mentoring, quickly agreed to journey on by herself.

The park superintendent and the Manager of the Concession both told her no person had ever made a botanical collection where she wanted to venture. Then they added they were "most anxious to have it done." Mexia realized this was an area of nearly 3,000 square miles and couldn't believe she was first.

Mexia later wrote Dr. B. L. Robinson, curator of Gray Herbarium, Harvard University (September 27, 1928) about her trip:

"My choice of McKinley Park proved to be a fortunate one; for I found that only local collecting had been done in the great interior valleys of Alaska, and not very much of that. Coastal Alaska, the islands, the Yukon, and around Dawson and White Horse have been repeatedly collected for many years, but until the Alaska R. R. was built there had been no way of penetrating the interior, and they don't know what they have up there. When I came to the Park, I was assured by the Superintendent and the Manager of the concession there that no person had ever made a botanical collection there, and that they were most anxious to have it done. At least we know that geographically no specimens have been secured from that locality, and as the Park comprises 2700 square miles it should make an interesting study. I hardly believe that there is another of our National Parks of which that could be said and I would like to carry it on."

Durlynn Anema, Ph.D.

Transportation became her greatest problem. The railroad stopped at the park, so it was up to Mexia to find a way into the wilderness. Part of the time she hiked, using a few pack dogs to carry her equipment. Other times she carried or hauled the equipment by herself.

Nothing deterred her. The park rangers assured her they would come to her campsites occasionally, but never showed up. She did have some visits from hunters, and even a group of backpacking tourists. But most of the time she was by herself with only her dogs for company.

She faced difficult collecting and hiking conditions daily. "I could only scratch the surface of that vast Park area, but I chose my stations as far as possible from each other and at different altitudes."

Her collection route spanned almost one hundred miles, beginning at the Savage River, moving to Copper Mountain (as close as she could get to Mt. McKinley), and on to Wonder Lake, a lower elevation that helped her avoid the early onset of snow.

Savage River, at the upper edge of timber line, twelve miles from the railroad and the entrance to the park, was her first collection area. Mexia easily hiked this distance in a day, carrying her equipment so she could remain for several days. Her pack dogs, "a novel method of transportation," worked tirelessly hauling her driers and other equipment.

Collecting specimens proved far more difficult than she anticipated. The roots of the more fragile plants broke off when she tried to dig them up. They were embedded in lichens and mosses. Below the lichens and mosses was an intricate combination of fine woody roots and the underground pliable stems of willow trees.

Mexia tried using a botanical pick. But the pick, she later wrote, "made no impression on this elastic 'wire mattress' sort of growth, for one never seemed to reach real earth, and each plant had to be dug out with a knife, cutting away these roots, and generally getting broken in the process of extraction. I am hardly likely to be up against just that condition again and so hope for better roots."

THE PERFECT SPECIMEN

> Mexia later wrote a complete description of her problems while collecting to Dr. Frances W. Pennell, Curator of Plants, Academy of Natural Sciences, Philadelphia December 10, 1928):
> I could only scratch the surface of that vast park area, but I chose my stations as far as possible from each other and at different altitudes. The first was 12 miles within the entrance of the park at Savage River, at upper edge of timber line. The second was at Copper Mtn.(sic) 60 miles into the mountains and as near as is practical to get to Mt. McKinley. This was a three day pack train trip in. I stayed alone up at Copper Mtn, which is at the upper margin of brush line for some three weeks, and then went to Wonder Lake, dropping down somewhat as snow was beginning to fall higher up. Wonder Lake was 22 miles further, near the S. W. boundary of the park, and the only way to get there was to hike, carrying what one could, and assisted by three pack dogs that carried some of my driers. Certainly a novel method of transportation for me.

The weather presented a problem throughout the expedition. Cold rain and wind often prevented her from collecting. Further hindered by the rough terrain, Mexia was concerned that she "only was able to get in the neighborhood of four hundred numbers, possibly not that many."

She wanted to stay longer. However, when it began to snow heavily, she knew she had to leave. Confused by the deep snow, she did not know which way to go so hesitated to move. Finally on September 12, 1928, she and her dogs were rescued from her campsite by an Alaskan on a dogsled.

As Mexia looked back at her trip she realized the difficulties she faced. In a letter sent to Dr. B. L. Robinson, Curator, Gray Herbarium, Harvard University, she described the Alaskan conditions:

"My trip to Alaska lasted nearly four months, and I only returned then because the advancing season cut off my supply of plants. I found the country and the flora most interesting, working most of the time between timber and brush line or on the upper edge of the brush line and as far up the slopes as I could get. There was abundance of material, but the weather was inclement and it was hard to get the specimens dry. Transportation, when away from the railroad is a real problem in Alaska, and part of the time it was hiking and packing with the assistance of a few pack dogs, and one's own back.

"My choice of McKinley Park proved to be a fortunate one; for I found that only local collecting had been done in the great interior valleys of Alaska, and not very much of that. Coastal Alaska, the islands, the Yukon, and around Dawson and White Horse have been repeatedly collected for many years, but until the Alaska R. R. was built there had been no way of penetrating the interior, and they don't know what they have up there...

"...The difficulties of the country and the climatic conditions were great so I was only able to get in the neighborhood of four hundred numbers, possibly not that many. My collection has not yet arrived, but I presume will come through safely. It will probably not be ready to deliver until early in 1929. Will you be interested in a set?"

(Mexia here displayed her business acumen acquired years before in Mexico.)

The Perfect Specimen

When she returned she began working first to finish her Mexican collection and sets with Nina Floy Bracelin, a recent botany graduate who had agreed to help her. When Mexia had left for Alaska she had not yet finished her Mexican collection. Completing her collections was a big problem for her. While she loved to collect, she was not excited about working diligently to quickly finish a collection so it could be sent.

However, while Mexia was gone Bracelin had been hard at work on the collection. Her diligence meant that by September 27, 1928, Mexia was able to send the collection to Dr. B. L. Robinson, Curator, Gray Herbarium, Harvard University, along with this letter:

"I am today mailing you the balance of your set of Mexican plants in six packages containing 782 specimens. These, with the previous shipment of 216 specimens, brings the total number to 998. I am enclosing the bill.

"The specimens have been very carefully packed this time and I hope will reach you in good condition. I have been extremely sorry that the ones sent earlier suffered so in transit, and as I have some extra duplicates, I would be glad to replace any injured specimens that I still happen to have if you send me the numbers."

Her choice of Bracelin as her assistant was a wise decision, and began a relationship which would last the rest of Mexia's life.

One set of her Alaskan collection was to go to the Academy of Natural Sciences in Philadelphia. Although she and Bracelin were working on the set, evidently it was not yet complete in December 1928. She wrote to Dr. Frances W. Pennell, Curator of Plants at the Academy, explaining the delay.

"I have had to interrupt work on the sets to get out for Mr. Ansel Hall a list and descriptions of some of the commoner species for the park pamphlet for McKinley, as this is going to press. The specimens will not be ready for delivery for some little time yet but I will make up your set if you so desire, taking special care that you get every Scroph I found. They and the Saxifrages were the most abundant."

Throughout 1929 her Alaskan sets finally were sent to all Herbariums who had supported her Alaska trip. She also sent sets to herbariums and botanists who had heard of her Alaskan botanical work.

Dr. Pennell sent confirmation of receiving the set and complimented Mexia on the excellent condition. Mexia replied on September 30, 1929: "I am very glad that you liked the Alaskan plants with which I took much pains."

Alaska was a challenge for Mexia, one she was glad she had accomplished. However, she knew that the plants of the Southern Hemisphere most intrigued her. Now she would make plans for more extensive research in South America.

Chapter Nine

Becoming Well Known

Mexia's changes in fifteen years often amazed her as much as they satisfied Dr. Brown. He had sensed there was another person behind the insecure, frightened woman who first came to him in 1909, so was thrilled he had unleashed a woman with potential to influence the future. For him, this gave total satisfaction.

Life was far different for Mexia than when she lived in Mexico: now she had a life with a wide collage of friends, activities, and successful career. She regularly kept in contact with Dr. Brown and his wife because he not only was one of her dearest friends, but mentor and confident. She saw his pride in her botanical accomplishments -- so made sure to tell him her adventures after each exploration as well as sending letters during the trip.

Mexia also had a wide range of friends in several disciplines. The Sierra Club was always on her list. Whenever she returned from an exploration she would be sure to attend meetings to tell her latest activities. Dear to her heart was the Save-the-Redwoods League whose headquarters was in San Francisco. She would visit their office, tell of her latest discovery, and how different collecting in Mexico's tropics was from California. These organizations comprised many of her friends along with

Durlynn Anema, Ph.D.

Alice Eastwood and the botanists of the California Academy of Sciences.

These friendships constantly thrilled Mexia as she recalled how terrified of people she had been in the past. Now she felt warmth in their presence.

However, one friendship had come into her life which thrilled her the most.

In 1927, Mexia met Nina Floy "Bracie" Bracelin when they both were enrolled in a University of California extension course. Bracelin had arrived at the University of California, Berkeley as a student. She was born in 1890 in Star Lake, Minnesota and had private tutors prior to being enrolled at the university. Fascinated with botany, she became a researcher at the university's herbarium after graduation.

Mexia and Bracelin enjoyed being together from the beginning, becoming good friends because of their mutual interest in botany. The younger woman looked up to Mexia and her expertise in collecting, was fascinated by the places she went, and the dangers she faced. However, Bracelin was much more interested in assisting with the collections than making those trips. Going into the wilds and camping for days did not appeal to her. Assisting Mexia in sorting out the collections, and sending them to the proper institutions did appeal to her. This proved to be a valuable relationship.

When Bracelin agreed to assist Mexia, she discovered immediately that while the collections were vast, they also were very disorganized. She never said anything about this to Mexia, just mentioned that she loved to organize and would be happy to do those tasks while Mexia collected. Mexia was overwhelmed by this offer, quickly agreeing to their relationship. Now she would be free to roam and collect without having to worry about how herbariums would receive collections.

Starting in 1928, Bracelin labelled specimens and sent them to experts for identification, developing a large network of corespondents who were happy to work with her. She started with the Mexican collection Mexia had amassed on her trip in 1926/1927. Mexia had been working on this collection since she had returned, but with 33,000 specimens it was overwhelming.

THE PERFECT SPECIMEN

By the time she returned from Alaska, the job almost was complete and ready to send by late September thanks to Bracelin.

> John Thomas Howell, Curator Emeritus, California Academy of Sciences, mentioned in a letter that "To facilitate the handling of the large collections of plants that came from her various expeditions to Mexico and South America, Mrs. Mexia engaged the help of her friend N. Floy (Mrs. H. P.) Bracelin, who at one time was an assistant in the University of California Herbarium."

There is some speculation about the relationship between Mexia and Bracelin. Both of Mexia's marriages failed due to her lack of understanding about intimacy, and the true relationship between a man and woman in marriage. A new relationship with a man never entered her mind because she had new interests concerning the environment and botany. Bracelin evidently had been married at one point in her life because in some references she is referred to as "Mrs. H. P." Obviously, the marriage did not last. During ensuing years and at Mexia's death, Bracelin talked of Mexia as a mentor and a mother figure. Whether theirs was a mother/daughter relationship because of the twenty year difference in their ages -- or something more is not known.

Bracelin remained one of Mexia's closest companions until her death, despite their twenty-year age difference. Whatever the nature of their relationship, it was positive for both of them. Mexia was free to devote more time to the field, and Bracelin received a steady stream of specimens to sort, process, and complete.

Bracelin had the specimens identified, mainly by the Gray Herbarium at Harvard University. She sorted them into sets and sent them to the various subscribing institutions. She dealt with paperwork, raised funds, and wrote correspondence. Bracelin also transcribed many of Mexia's letters, recognizing the importance they would have in the future. This left Mexia time to plan her itineraries, give presentations, and decide her next trips.

Bracelin first worked on Mexia's Mexican collection at the University of California Herbarium at Berkeley because that was where Mexia had arranged for the specimens to be delivered. In May 1929, Dr. E. B. Copeland, curator of the Herbarium, hired her as an assistant. In the early 1930's, Bracelin helped Dr. Carleton R. Ball in working up his willow collections for the revision of the genus Saliz in the western United States.

Later she made a collection of the exotic plants growing in the Anson and Anita Balke estate (which became part of the University of California, Berkeley). With duplicates, her 1,392 garden collections amounted to about 20,000 sheets, all of which were distributed to herbaria expressing an interest in cultivated plants.

After leaving the University of California Herbarium, she perfected her skills as a scientific illustrator (specializing in the field of graphs and charts). From January 1940 to July 1943, she was an assistant in the Botany Department of the California Academy of Sciences, San Francisco. After that and until her retirement in 1960, she was on the staff of the Western Region Research Laboratory, U. S. D. A., Albany, California.

She also wrote an article for *Madrono* titled "Itinerary of Ynes Mexia in South America."

Bracelin not only processed Mexia's collections, but made a name for herself in the Botanical Field. This information was printed in the "Memorials" section of *Madrono* in 1973 following her death on July 8, 1973. Author Annetta M. Carter, Department of Botany, University of California Berkeley said about Bracelin:

"She is remembered as a cheerful, friendly person, ever helpful to others, with a great capacity for work that was well done."

Carter also commented that "Except for the dedicated and meticulous assistance of Mrs. Bracelin, the extensive Mexican and South American collections by Ynes Mexia might never have been distributed."

The Perfect Specimen

In the San Francisco Bay Area, Mexia often was asked to present about her trips, explorations, and discoveries. Her travels were well known through newspaper articles appearing whenever she returned from an expedition, resulting in several speaking invitations as a popular lecturer. This shy "little girl" had become so confident in her abilities, she was able to entertain audiences with her tales, photographs of her travels, and collecting.

> An example of Mexia's lectures was a series of free illustrated lectures presented by the California Academy of Sciences on the subject: The Beauties of Nature in the Fall of 1932.
> Mexia's lecture was on October 19.
> "Up the Amazon and Over the Andes:" An account of a trip for the collection of botanical specimens secured for the California Academy of Sciences, the University of California and other institutions. The topography and geography of the Amazon Basin will be described and illustrated with numerous lantern slides. The lecture will be delivered by Ynes Mexia who, after a stay of two and one-half-years in Brazil and on the East Coast of South America, crossed the continent at its widest point from east to west.

She also wrote several articles for the *Sierra Club Bulletin* and *Madrono*, the journal of the California Botanical Society. (See bibliography for the titles.)

One her favorite speaking engagements was for the Save-the-Redwoods League. She had a symbiotic relationship with the League and was convinced the more she spoke and made people aware of the dangers to these magnificent trees the better it would be. In a June 10, 1927 memorandum to Mexia from the Office of Save-the-Redwoods League they sent her suggestions for a four-minute talk on the Redwoods and the Movement.

Durlynn Anema, Ph.D.

The conclusion of the Memorandum said: "The League appreciates your desire to 'spread the gospel' of the Redwoods by this talk, and we hope that the material sent you will serve your needs."

Mexia's friendship with Alice Eastwood also continued. They went on several excursions together whenever she was in the San Francisco Bay Area. Mexia always was delighted to accompany this renown scientist, grateful not only for all Eastwood had taught her but for her willingness to buy what she collected.

John Thomas Howell, Curator Emeritus of California Academy of Sciences, relates one such trip when he and Eastwood were collecting.

"Mrs. Mexia was a close friend of Alice Eastwood ... and in 1933 accompanied (chaperoned?!) Miss Eastwood and me on the first 'Eastwood and Howell' collecting expedition. It was quite a trip in an open Model T Ford that traversed parts of Nevada, Utah, Arizona, and California between June 6 and July 3 and netted over 1300 collection numbers. While Miss Eastwood and I collected plants along the way, Mrs. Mexia took pictures -- she was an excellent amateur photographer. When we reached the Grand Canyon North Rim, Mrs. Mexia and I spent June 23 going down into the Canyon to Roaring Springs; she on horseback, I on foot. After lunching together beneath the cataract at Roaring Springs, she relieved me of my morning's collecting, carrying that part of the day's haul back to the rim on her horse. Thus relieved I was able to collect further on the 4000-foot climb out of the canyon. It was a superlative day, netting over 100 kinds of plants, one of which, *Carex curatorum*, was new to science. The printed labels for that day should rightfully have carried Ynez Mexia's name as well as those of Eastwood and Howell.

THE PERFECT SPECIMEN

While Mexia spent the next months working on her collections, she was still anxious to return to the field. She took a minor trip to Mexico in 1929 for three months, visiting the states of Chihuahua, Puebla, and Hidalgo, and collecting over 5,000 specimens. The trip also allowed her to settle some final business issues related to the ranch she had once owned. Her life had changed greatly over twenty years. She'd gone from being a wife situated in one location and essentially running a business, to a globe-trotter who rarely slept in the same place for more than a short period of time.

Now she was ready for the major trip of her career -- going to the far reaches of Brazil and beyond.

Mexia's practical outfit for exploration and collection.
*Courtesy of California Academy of Science,
San Francisco*

CHAPTER TEN

EXPLORING BRAZIL FOR A YEAR

"I have come up against a snag...Can you get me officially appointed as your Assistant?" Mexia frantically wrote Agnes Chase regarding getting a visa for Brazil. She was scheduled to join Chase, an associate agrostologist (studies grasses) from the Smithsonian Institution, on an expedition west of Rio de Janeiro in mid-October 1929.

Mexia was excited about this new adventure because South America was unfamiliar territory. The challenge of an unfamiliar continent could be intimidating to even the most experienced traveler, but for the solitude-loving woman it was exactly what she wanted.

Now her visa problems threatened to ground her indefinitely. She realized officials in Washington, D. C., might expedite the process, thus her request to Chase to be appointed as an official assistant to the expedition. This appointment would give her more authority in dealing with the government. In return for the favor, Mexia promised to collect double specimens and send one set to Chase.

The visa problem resulted from several factors. First, there was no Brazilian Consul in San Francisco, so Mexia had to send her request to either Chicago or New York. A Brazilian agent told her it could take up to a month or two. Secondly, when she married Mexican citizens she lost her citizenship, which she regained in 1924. However, she lost the certificate of restored citizenship so had to send a petition to Washington, D. C. for her birth certificate. This was a slow process and hence her request for the appointment.

Fortunately, Mexia's pleas were answered in time. On October 15, 1929, she left San Francisco for San Pedro, California where she would board a Norwegian freighter bound for Rio de Janeiro. She had letters of introduction to officials throughout Brazil including the Amazon region. Aware that the populace of Brazil spoke Portuguese, and not Spanish like the rest of Central and South America, she hoped her Spanish would help her translate. This was her first trip among people in whose language she wasn't fluent, and she saw it as an opportunity to learn some Portuguese.

Her departure date was fortunate. Several days later, the stock market crashed, sending the country into the Great Depression. Any later and her trip might have been cancelled.

What Mexia and the sixteen other passengers had not anticipated was the lack of staterooms -- only twelve staterooms for seventeen passengers. This was the first time the freighter had taken passengers aboard -- a learning experience for both crew and clientele. Also, there was only one bath and other facilities for all the passengers and the officers. Her roommate was a "very pleasant" elderly woman from Bolivia, assigned to her because she spoke Spanish.

When Mexia learned another female passenger had a canvas hammock on the upper deck, she decided she also would be in the cool night air. "I unpack my camp cot and camp out on the other side of that deck at night, and am cool and happy," she later wrote. Because the boat did not stop at any ports, the passengers were bored. Not Mexia. She had brought along enough reading material to keep her occupied during the month long trip.

As the freighter journeyed south, Mexia also relaxed and enjoyed the bracing ocean air. She later wrote, "The trip down to the Canal is a pleasant one; some light rains and all the way down a blessed breeze that keeps us from becoming uncomfortably hot. The sea gets bluer and bluer and the whale is

The Perfect Specimen

spied in the distance. Land birds are with us, a small owl that looks much like a burrowing owl, a meadow lark and some small sparrow-like birds come to us from the just visible shore of Lower California."

The freighter reached the Panama Canal on October 29. Mexia not only was impressed by the journey through the locks, but also at the fascinating scenery. "There was so much to see that I was kept scampering from one end of the boat to the other. It was such a relief to see land with real vegetation and trees on it after nearly two weeks. Going through the Canal is the apparent ease and simplicity of the whole proceeding. One goes in between walls of masonry, nothing extraordinary in appearance, and rests a while and some parts of the masonry apparently detach themselves slowly and close and you see they are the great lock gates."

Once out of the canal, the freighter sped toward the eastern coast of South America, crossing the equator. Mexia thoroughly enjoyed the traditional ceremony for passengers and crew who had never crossed the equator. It took place at the ship's small swimming pool made of canvas and wood, where she swam at least two times a day. People were lathered with paint and soapsuds, then thrown into the pool.

Mexia decided to have a little fun with the ritual: "An imp of perversity prompted me and I thought I would give them a run for their money." She went limp as she was thrown into the water and let herself sink to the bottom. When the scared guards pulled her out and propped her against the side of the pool, she slipped away again. As she pretended to drown, her head went under "as any properly drowned person does." The guards were afraid to grab her, yet more afraid not to rescue her limp body. After they pulled her out, they hesitated again, looking for help. But the rest of the crew was equally scared.

Mexia commented later, "I wanted to suggest artificial respiration but it did not seem to come properly from the 'body' and all the rest were too frightened to think of anything, so after a while I had to come to all by myself without any artificial anything at all, but the joke was not on me this time as it had been with the others."

The bashful little girl she once was had disappeared. She could talk with everyone, felt much more comfortable around people, and certainly enjoyed playing a joke. Consequently, Mexia enjoyed the trip, especially walking the available deck space and talking with her fellow passengers. The monthlong trip was restful, as she wrote at the end of the letter: "... it has given me time to catch my breath before tackling America del Sur."

Chase and the rest of the scientific team met Mexia in Rio de Janeiro and explained their plans. Their ultimate destination was Espiritu Santo, where they would travel into the foothills of Pico de Caparao, the highest peak of the range. This site was chosen because, to their knowledge, no botanizing had been done on the southeastern side.

The group traveled by train to Alegre. After transferring all the baggage and equipment into a truck, the group climbed in and experienced a muddy, rough, and winding road trip. As they climbed, wonderful mountain views appeared as well as unknown vegetation. Many steep slopes were cultivated with coffee bushes. Others had been denuded to plant corn and beans, but now held only weeds. Growing naturally among this agriculture were cecropia trees, "perhaps the most conspicuous, for they have tall slender trunks with a great crown of huge palmately incised leaves that stand out in a perfect crown showing silvery white on their glistening surface."

The trip proved to be the roughest Mexia had ever undertaken. Mexia claimed she was unprepared because she "had no idea what it was going to be." She found Chase, being a "pure and simple" scientist, was "unpractical as all scientists are." Chase would constantly say, "we will get along somehow," even protesting when Mexia insisted on taking food for a journey into the wilderness. But Mexia was firm, saying later "or we would have starved as well as nearly frozen."

On their first day, the group climbed through the rain forest to Pico de Caparao's 9,000 foot peak. There was little available in the way of beaten trails. The constant mist and rain meant all the foliage dripped. The group was drenched, and had given up any hope of trying to stay dry.

The Perfect Specimen

On the first night, their guide wanted to stay under some overhanging rocks, but Mexia found the dry earth alive with fleas. She was able to convince the guide and porters to level some ground beyond this place and set up a rough tent of a waterproof alligator skin and Mexia's poncho. At that point, she discovered Chase had not even brought a change of clothing or warm outer coat, and only a thin blanket. Mexia shared her clothes while growing frustrated at the woman's lack of common sense.

The next day, they continued to struggle upward through the "green gloom" of the forest. While Mexia found the forest magnificent and the ferns so lush they "would have turned a fern herbarium green with envy," they could not think of collecting, but simply plowing forward. Chase's knees became wobbly and by late afternoon Mexia was so short of breath it was difficult to keep climbing. Stops became more frequent. Their guide said it wasn't much further to the timberline. They finally moved out of the taller fronds into a bamboo thicket and fought their way through the bamboo tangles until they cut through the "green earth drapery" and were above the timberline.

Chase and Mexia were relieved but very tired. A cold wind sped down from the mountain peaks and over the soaked, bedraggled women. Chase's teeth chattered and Mexia, to her shame, became nauseated. They sat at 8,000 feet on the tallest mountain in Brazil and wondered how it could be so cold when they were so close to the equator. The guide and porters wanted to go to a hut five miles further on, but the women were too exhausted to continue.

They insisted on stopping and talked the men into again erecting the tent after leveling off the space and digging ditches for water drainage. The men built a fire, then went on to the hut. Mexia and Chase sat inside their makeshift enclosure and ate bread, cheese, and part of a chicken before lying down for the night, listening to the steady rain and howling wind.

When they awoke, it was still raining. After eating some cold provisions and not knowing where the men's hut was, they decided, because they already were so wet, they might as well go out and collect. They found so many specimens they became too involved to notice the time. In the early afternoon, two of the men came down from the hut to try to convince Mexia and Chase to press on. The women declined. They had collected so much that they wanted to put the plants into the press, making them easier to carry.

Chase was much speedier than Mexia in collecting grasses so helped with the pressing and packing. As they worked, the storm grew more intense. They worked until it became so dark they had to stop. They crawled into their damp bedding and went to sleep. Then the wind blew down the poncho. Mexia knew she had to fix it so "stripped and pranced out in the rain and dark to fasten the end up as best I could. It really is not as cold without clothes as in wet ones and after the first shock I thought the shower quite a lark. I needed it!"

The men returned the next day. By this time, everyone in the entourage was ready to go to the other side of the range -- perhaps to find sun. On the other side, they found a Brazilian family who ran cattle in the hills. The family took the sopping wet group in, gave them hot water, a hot supper, and put them up for the night. The next morning the sun was out, bright and clear. Mexia and Chase spent that day drying out their belongings and changing the wet papers on their plants.

The next day Chase woke up complaining about being cold all night. She had a severe sore throat, so wanted to get back as soon as possible. The guides also felt this way before the rivers rose. The downhill trek was comparatively dry, so the women could continue to collect. Chase, although becoming even more ill, insisted on helping Mexia because there were few grasses to collect.

By the time they reached the village of Santa Barbara de Caparao the cold, wet weather had taken its toll on Chase. She was very sick upon arrival, but insisted staying one more day to help Mexia who wrote later: "She is an indefatigable worker and very quick, while you know I am slow." Mexia greatly appreciated Chase's stamina and determination in face of illness, but emphatically told her to return to Rio immediately, which she did the next day.

The Perfect Specimen

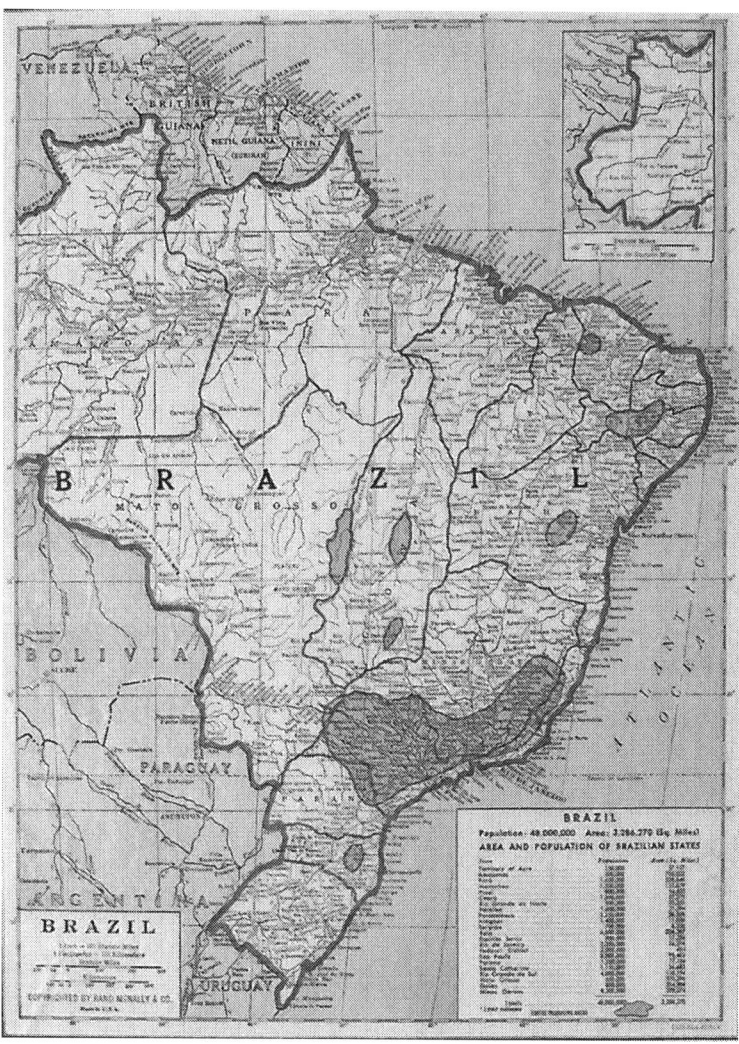

Mexia's first year in Brazil was concentrated on Minas Geraes.
Map compliments of the National Coffee Department of Brazil.

Mexia stayed behind to collect more plants, then went on to Rio to plan her next move.

By early December she was in the state of Minas Geraes which she discovered was much larger than Texas with a million more inhabitants. "It is a fact that Brazil is so much more thickly inhabited than one suspected that strikes me I think. Everywhere, except in the rain forest on the flanks of Caparao among the tangle of vegetation are little scattered mud huts even in what appear the wildest places."

Mexia went to the town of Vicosa to visit the Agricultural College, and found the school a perfect location for collecting. She was impressed to find such a modern institution so far from what she considered civilization: "Out of the, luckily fertile, red clay soil here they have evolved a truly modern institution, with buildings, out buildings, and planting that would do credit to any country in any locality."

The school had been founded twelve years previously. There were over one hundred students, all male, and all native Brazilians. They learned about agriculture and received hands-on experience with cattle breeding, feeding experiments, and crop improvement. The cattle were large, colored cream to pale gray, with wide-spreading horns and the characteristic hump and pendant dewlap of the zebu. This type of cattle is raised in Brazil because they are useful in both pulling carts and plowing. These cattle can live on dry grass in the dry season and withstand Brazil's very hot weather.

Mina Geraes became Mexia's home for the next year. She divided her time between Vicosa, a fazenda (ranch house) in the Sierra da Gramma, and another fazenda north and west of Vicosa. These collection localities were in the highlands of mid-Brazil at an elevation of 2,000 feet. This was Mexia's longest journey away from the United States, and she had no intention of returning any time soon.

Her main collecting was done in an area three days ride from Vicosa. According to Mexia, "this locality had been selected because it is over the ridge that forms the watershed in this district, and as it faces the east, it gets the precipitation necessary

The Perfect Specimen

to make High Rainforest, hence the collecting should be good -- especially of the Pteridophyta and the Epiphytes." Dr. Rolfs and Mrs.Chase had climbed Serra da Gramma five years previously and were impressed with the collecting possibilities.

Mexia benefited greatly from having a local student assistant, Joaquin Braga, who was as interested in botany as she was. Her expedition's caravan consisted of three pack mules, each with two large fiber cases of driers and other collecting apparatus. Three other saddle mules carried Mexia, Braga, and the packer.

Mexia was approaching sixty years of age, but she rarely slowed down. She was amused by the reaction of villagers when they saw her ride onto a main street: "The village was treated to the sensation of all time to see a woman astride. I feared knickers would be too much for the Brazilians, but even a divided skirt was beyond their wildest imagining, and every door and window was filled with, I suppose, shocked spectators."

She found a joy in the riding part of her trip because she could go along slowly enough to see the scenery and observe the vegetation. "In collecting often one cannot 'see the woods for the trees' literally, and while an occasional novelty arouses the collecting urge, in general I am content to enjoy my surroundings as I ride along. It is almost impossible to collect on a pack train trip, for one cannot hold up the heavily loaded animals. And then the material has to be cared for; the collecting is the least part of it."

She still preferred to sleep outdoors to avoid the fleas that seemed to inhabit every house. Her hosts were often shocked by her choice, fearful she'd be attacked by a ferocious beast. Mexia found it delightful.

> "To be in a rainforest in Brazil was thrill sufficient," she wrote to friends. "It being a high rainforest, one thousand to one thousand three hundred meters, it is not so dense that one cannot see into it, and the great trees, many with lianas equaling their girth, stood above everything. Then the tree ferns are really the most beautiful and graceful things ever created; the stems are slender, showing their leaf scars, and then from the top the great fronds arch out in most exquisite curves and each is lacy perfect. One thing I was surprised to find ... ferns are always so smooth and unharming, and the great petioles of these ferns as thick nearly as my wrist, had very decided prickles in the form of recurved hooks not to be trifled with. Why should a fern have prickles?"

Part of the importance of botanical collecting is the insight it can give into how living things exist in the wild and adapt to their conditions. Just as polar bears are white to match their snowy environments, or fish come to resemble the river bottoms they inhabit for protection, so also do plants adjust to ensure their survival.

Mexia's year in Minas Geraes ended with a bang. One night, as she, Braga, and the rest of the porters slept under the stars, a monstrous thunder and lightning storm set upon them. Her team was shaken with the experience.

The next morning Mexia suggested building a shelter for the men, but they only wanted her to finish collecting so they could return to the fazenda. Braga also had been scared and suggested they collect quickly, and then start home that afternoon. Mexia sent the two porters and the assistant to collect a special plant while Braga packed. Although it was not raining, it was cold, windy, and cloudy. Mexia didn't get all the varieties she wanted but did have the press full by eleven o'clock. They were ready to leave by noon.

The group started down the mountain -- straight down. Mexia couldn't decide whether this was by chance or by design. They were descending over steep rocky, waterworn granite. They had to creep along the edges or take advantage of cracks when they appeared.

Mexia later discovered the porters had been so scared by the storm they wanted to get home as quickly as possible. Thus, they took the steep, hard trail rather than the easier one by which they had ascended. Mexia was disappointed, not because of the difficulty of the descent, but because she could have collected more plants on the first trail.

After five hours, they reached the level part of the trail and had a smooth hike the rest of the way to the fazenda. While she wished she could have collected more, Mexia realized that aside from the grasses collected by Chase, no collecting had been done in this area. She hoped her collections would encourage more exploration.

A truck took Mexia, her equipment, and baggage to the second fazenda. Braga returned to the Agricultural College. Her stay was on a large fazenda on which they raised cattle. A housekeeper/cook took care of her every need. The fazenda also raised enough coffee, rice, beans, sugar cane, and pork to be self-sufficient. She was appalled that the vegetables were never eaten, and fruit only when it could be found. Fifty cows provided milk both for the fazenda and for a commercial creamery.

Mexia was fascinated with the Brazilian Cattle Ranch and later wrote about it in "Glimpses of a Brazilian Cattle Ranch," which was never published.

Her view:

"Brazil is a large breeder of cattle for beef and hides, and there exist huge 'Fazendas' (Portuguese for Haciendas or ranches) where cattle raising is carried out on a large scale. ... In the state of Minas de Geraes I visited the Fazenda do Diamante. This Fazenda stretched for leagues and leagues, all under a delimiting fence with miles of wire fence dividing the pastures. Ten thousand head of cattle were here with special pastures for the calving cows, the yearlings, the heifers, and the steers.

Durlynn Anema, Ph.D.

Brazilian cattle.

Mexia was fascinated by the cattle ranches she visited in Brazil. *Photo Courtesy of Durlynn Anema*

Mexia also was excited to find rheas out in the open, but mistakenly called them "emus" (both are relatives of the ostrich): "One day I actually had the thrill of seeing three of the great birds at a distance. They were running, having probably sighted us first, but even far away they looked huge, and were a sight not to be forgotten."

While she could have stayed indefinitely in the highlands, Mexia felt the need to follow her original dream -- to travel the breadth of South America along the Amazon. So she said goodbye to new friends and the beauty of the area, and went to Rio to catch a boat to the Brazilian state of Para and the beginning of another adventure.

Chapter Eleven

Up the Amazon

The botanical treasures of the Amazon River, especially at its origin and tributaries, drew Mexia like a magnet. She had dreamed of visiting this area for years: "Most of us, I think, have felt the fascination of the Amazon region. So much have we heard of its rivers, its tropical beauty, its luxuriant forest, the wild life and wilder Indians that lurk in its depths, that the pictures drawn by our imagination are vivid and unique. This vision of the unspoiled wilderness drew me irresistibly."

Mexia spent almost twenty-two months in Brazil before she was able to accomplish her dream. Because she was on the east side of the Andes Mountains, she decided to reach the Amazon and its tributaries from that direction. Immediately, she found that trying to obtain information about that route proved difficult.

Ways to Approach Exploration of the Far Amazon Area

As Mexia tried to find how to explore the Amazon and its tributaries traveling east to west, she was amazed at the answers.

One source recommended to go south to Buenos Aires, take the Transandean railroad to Chile, go up the west coast to Lima, and thence it was considered possible, though difficult, to traverse the Andes and descend to the Amazon, thus crossing the continent from west to east.

> Mexia also heard a fascinating tale that told of the remote gorge of the Amazon, the Pongo de Manseriche, where the river breaks through the Andean chain. (It does not.)
>
> Because she already was on the eastern coast, she felt this seemed a roundabout way to go because she already was on the eastern coast. However, her advisor had no information whether an east-west crossing of South America would be possible; he had never heard that it had been done. Later she discovered that ascending the great rivers and climbing the forest-choked eastern slopes of the mountains is a much more difficult proposition than the west to east passage.
>
> Few explorers had accomplished this east to west journey. Harriet Chalmers Adams in the early 1900's had traveled a small distance into the Amazon basin from the west across the Andes. After her explorations, she had returned by the same route -- this time east to west -- but it was a very arduous route. Also, her exploration was only a minute part of the total basin.
>
> Today the journey usually is by airplane -- Lima,Peru to Iquitos, Peru.

The lack of information did not stop Mexia. She decided she would go as far as she was able.

After a few months of side excursions around the north coast of Rio, Mexia booked passage on a comfortable motor ship that stopped at all the ports going up the coast. Mexia was thrilled because it gave her a chance to watch the gradual change in Brazilian environmental conditions and modes of living from south to north. She also took the opportunity to visit the historic towns.

Twelve days later she arrived in Belem in the state of Para which was just within the mouth of the Amazon. This old city had beautiful shade trees and gardens with "many-hued" flowers.

THE PERFECT SPECIMEN

She was surprised that even though near the equator, the heat was not as oppressive as she expected. She also was happy to note that mosquito eradication had been so thorough she encountered none of the "pesky" insects.

Officials of the Museo Goeldi were presented letters of recommendation and gave her assistance in preparing for her latest adventure. She obtained visas for Columbia, Ecuador, Peru, and Bolivia, because she wasn't sure exactly where she was going. While she awaited her steamer she went to a newly established Japanese Agricultural Colony. She also did some collecting because the virgin forest was being cleared and it would give her a chance at some of the magnificent tree growth of the lower Amazon Valley. Collecting in the Para area itself had already been saturated.

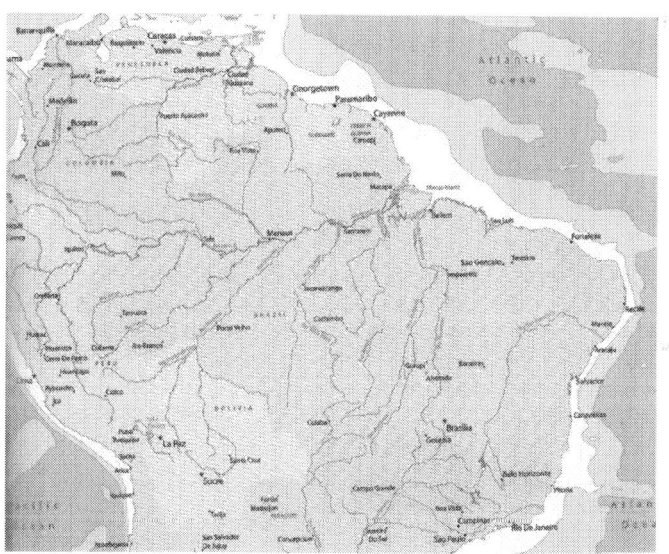

The Amazon River Basin is comprised of a network of rivers that span the northern portion of South America.

Durlynn Anema, Ph.D.

On August 28, 1931, Mexia boarded the steamer *Victoria* for the trip upriver. She was loaded down with equipment in preparation for a long stay in the region.

The steamer was an upgrade in the living conditions she always experienced while collecting. Mexia enjoyed air cooled by electric fans and ate fresh meat --"off the hoof" -- on this leg of her Amazonian adventure. "... comfortable river steamer built for the tropics, with airy, screened staterooms, electric lights, ice-plant, etc. Meals were served on the cool, unenclosed deck amidships."

As she sat on the deck she absorbed the beauty of the river and its islands. These memories later became part of several articles she wrote. "The river itself is a tawny flood, looking more like an inland sea, 'El Rio Mar de las Amazonas,' than a river. Everywhere it is island-sown, and these islands divide it into parana (channels) each of which may be several miles wide. Vessels ascending the river follow these side channels, often bringing the boat sufficiently close to island-shore or mainland to enable one to see many interesting features. Every foot of terra firma is heavily wooded, and these forests of the Lower Amazon are truly magnificent."

The steamer was a wood-burner. Each day the boat tied up at some spot on the shore to obtain fuel to continue the journey. This gave Mexia a chance to go ashore. She was fascinated by the way the forest crowded the settlements of thatched houses, almost hemming them in, with little land left to cultivate. This was not the uninhabited wilderness she had imagined, because the steamer was rarely out of sight of a little settlement.

The *Victoria* reached Santarem, a good-sized town on the mouth of the Tapajoz River, on the fifth day. The huge rubber plantations of Henry Ford were several miles inland. This area had been visited for years by naturalist explorers, so she already was well acquainted with the vicinity. Most intriguing to her were the open-front stores hung with the skins of huge boas, lizards, caimans, and furs of monkey, ocelot, and many strange beasts.

The Perfect Specimen

On the trip's sixth day they came to a town called Obydos. This was the first time she could see both banks of the Amazon. Up to this point, islands had blocked the view. When they left the town, she began to see more wildlife -- huge jacares (caimans), white aigrette herons, and chattering green and silver parakeets.

Mexia enjoyed watching dugout canoes as their single paddler took produce or passengers up and down the river. Whenever the steamer passed a clearing with huts scattered about, all the inhabitants rushed to the shore and waved. The steamer's passengers always cheerfully waved back. Mexia marveled at the houses perched on stilts as a precaution against floods.

On September 2, 1931, the steamer entered the Brazilian state of Amazonas. It soon came to the mouth of the Rio Negro, where the boat turned north into the swift black waters. They arrived in Manaos two days later. This city had wide, tree-shaded streets, electric trams, a hospital, and other modern conveniences. It was renowned for its splendid public buildings and the beautiful opera house of Italian marble topped with a gold-tiled dome. This city in the middle of the jungle had developed because of the productive rubber industry.

Worldwide demand for rubber -- used in insulation and to make tires for the increasingly popular automobile -- was high. Before synthetic rubber was invented in 1930, rubber could be made only from the secretion of latex from rubber trees found in high concentration in the Amazon basin. In 1876, the British smuggled rubber-tree seeds out of Brazil, establishing plants in their colonies in Sri Lanka and other tropical regions. However, the Amazon basin still continued to be an important source of this valuable material.

> Manaos name was derived from the Manaos Indians, a warlike tribe who adorned themselves with grains of gold and bits of gold leaf, and who are supposed to have given rise to the many legends of "El Dorado de Manao" in which the Spanish Conquistadores believed implicitly.

When the steamer continued its journey, it carried a new group of passengers. Mexia was the only continuing passenger. The riverbanks changed, showing sand spits and newly formed islands. Near the water's edge was tall, coarse grass. Behind were imbaubas, fast growing trees that looked almost like palm trees, with slender silver-white trunks and enormous leaves that were covered with a down that made them gleam in the sun.

> In a later article, Mexia made some interesting comments about the people she observed along the Amazon.
>
> "As the steamer proceeds on its leisurely way deeper into the interior of the continent, the character of the people changes as well as the river scenery. The "caboclos" still show some negro blood, but more and more bronze skins take the place of black, hair becomes straighter, and features more angular as Indian blood predominates.
>
> The lives, even of the better-to-do people, are very simple -- a tiny clearing along some stream, a banana patch, a crop of mandioca, or cassava, perhaps a path of corn and beans grown on the rich silt left by the ebbing river; a palm-leaf house, a canoe or two are all they need."

As they approached the equator, Mexia braced herself for the anticipated heat, then was pleasantly surprised by the agreeable climate. Cooled by the moisture from the almost-daily rain, it was nothing like the scalding dry heat she had grown up with in the southwestern United States.

"Cafe-au-lait," which means coffee with milk in French, was Mexia's description of the Amazon waters. She loved eating the fresh fish, caught in abundance each day by men in canoes. The Amazon was very low because this was the end of the dry season. When the passengers heard a splash in the water, they would run to the side of the boat, often to see a huge slice of the tree-covered shore had fallen into the river.

The Perfect Specimen

On the twentieth day after leaving Belem, the steamer reached the Rio Javary, the river dividing Brazil and Peru. Mexia wrote that "it was with feelings of real regret that I left Brazilian soil. Despite my learned friend's prediction, the Brazilians had treated me everywhere with the greatest kindness and consideration."

Now they were on the Maranon River. The land to the south was part of Peru, and north part of Columbia. The Columbia portion once had belonged to Peru but now gave Columbia an outlet to the Amazon for the provinces lying south of the Andes.

At the steamer's first landing in Peru, Mexia was fascinated by the Iahuas Indians. Their dress consisted of a short skirt of split palm leaves, a cape, and armlets and anklets dyed an orange-red that shaded into their smooth brown skins: "Rather stunning they were, and quite willing to pose for their pictures in exchange for a few crackers."

On the twenty-fourth day, they arrived at Iquitos, Peru. "Iquitos is quite a lively town, sitting like a spider in the center of its web, whose silken strands are the shining rivers which come from north, west, and south, traversing this wilderness. The lanchas, or river boats, which ascend these rivers and their affluence, carry simple necessities to exchange for skins of beast, bird and snake, for rubber and mahogany, for vegetable ivory, and for monkeys and parrots."

Mexia finally had reached the end of her 2,500 mile journey up the Amazon.

Mexia stayed in Iquitos almost the entire month of October, preparing for her trip into the wilderness of the Amazon and its tributaries. She carried letters to several prominent people in the town, who were able to help her find lodging with a Peruvian family. Asking advice from several Peruvians, she learned that it was possible to continue the ascent of the Maranon River and enter the Pongo de Manseriche. The more she heard about the region's inaccessibility, its wildness, and its Indians, the more she wanted to go.

Mexia hired three men: Jose, who was half Peruvian and half German, as guide and hunter, and Valentino and Neptali as

Durlynn Anema, Ph.D.

cholos (porters) and canoe men. Food was supplied for the three months they planned to be gone. Mexia wrote, ".... quite different from our camp supplies -- farina or cassava flour in great baskets, "paiche," a salt fish and coarse brown sugar for the men with a very few tinned articles for myself." Trade goods for barter with the Indians also were included.

A lancha carried them up the Maranon River. It carried several passengers with Mexia in the "first class" section and her crew (she called them "her boys") in the third class section. She investigated their quarters and found them on an open lower deck where "freight, livestock and passengers were all mixed up together. The people sit around on the boxes or hang their hammocks anywhere they can find room."

It took a week to reach Barranca, where they then transferred to smaller craft. She and her hired men were "dumped ashore," according to Mexia. Then the riverboat "whistled thrice, turned and slid down the river."

In Barranca, Jose tried to hire a large dugout canoe. However, while every person seemed to have his own montaria (small canoe), large ones were not available.

Finally, he found one with four native paddlers, but it could only carry half the baggage. Mexia was tired of waiting, so she agreed to start up the river with Valentino and Neptali. Jose would follow with the rest of the baggage as soon as he could find another canoe.

The propulsion of the canoe fascinated Mexia. In the stern sat two of the paddlers who steered with their paddles. In the bow the other four men paddled with their round bladed paddles. The baggage was placed in the middle of the canoe. Mezia sat there under a small palm-leaf shelter and quickly forgot her rather "hard box-seat" as she watched the river and its life unfold. They crept along the river bank, often under overhanging trees, to avoid the current.

She relished the beauty around her: "The shining cream-brown river, stretching from sunrise to sunset, confined by living green walls on the right and on the left, and above all the high-arched sky, delicately clouded at dawn, its intense blue relieved as the sun rose higher by fleecy white clouds, which soon piled

The Perfect Specimen

aloft in huge cumuli, and turning black and threatening as they tore down upon us in a torrent of blinding rain, with thunder and lightning for the afternoon storm. The deluge lessened, passed us by, traveling Andes-ward, and left us crawling in its wake refreshed and enlivened under a cloudless sky until we headed into the burning heart of the tropical sunset."

Each evening they searched out a sandy beach to camp for the night. Valentino lit the fire and cooked. Neptali put up Mexia's cot and mosquito net, spread large Musa leaves for a rug, and then brought water for her bath. The river was too dangerous to bathe or swim in, due both to the currents and the fish and caimans -- a type of crocodile six to eight feet in length. While the men worked, she roamed around the campsite, watching the birds and examining the vegetation.

They arose at dawn to "inch the canoe" upriver, battling the heavy downstream current. Huge stranded trees stuck out from the banks. The current raged past their partially submerged branches. Occasional gravel bars between islands caused shallow rapids. The men had to be careful at all times.

One of Mexia's most vivid descriptions came after seeing the Andes for the first time: "One day, as we started westward, a blue mist hung low on the horizon athwart our river highway, which, unlike other morning mists, did not dissipate with the rising sun, but took on a dim outline and a deeper blue until it dawned upon us that it was no mist, but the eastern-flung chain of the mighty Andes, the barrier that would end our journey."

One day, Mexia and her companions looked up to find the riverbank lined with Aguaruna Indians holding copper-headed spears and twelve-foot blowguns with tiny darts quite visible. Mexia and her guides were startled and frightened. These natives assumed her canoe contained Wambisas, members of a native tribe who were their blood enemies.

"When they found we were 'Christianos' instead of the dreaded Wambisas, they were greatly relieved and received us with rejoicing," Mexia later wrote.

In preparation for meeting the various tribes along the river, Mexia had brought presents. She presented each woman with a needle and each man with a small fishhook as goodwill gifts.

Then the Indians took Mexia and her guides to the moluca (communal house). As in other locations, the natives were thrilled to have their pictures taken. She described them as wearing "a sort of skirt made from the wild cotton which they spin and weave, or from a fibrous bark beaten thin. The women had a kind of garment tied over one shoulder. The boys go naked."

Mexia's expedition had reached the point on the Maranon River where only canoes could go -- and these slowly, either creeping from rock to rock, or hauled up by ropes. When the river rose or was in flood, the rapids were immense in this narrow passage and no craft could traverse it. This was the famous Pongo de Manseriche.

They were fortunate to arrive as the river was falling, allowing them to move forward. Her paddlers were experienced river men, so they advanced easily in the dugout canoe. Mexia described the gorge as "gloomy" and dense with vegetation from top to bottom. The gorge was 330 feet deep, too deep for rapids. However, the river moved rapidly from side to side in the narrow canyon, forming ferocious whirlpools. The water welled up in standing waves and rushing crosscurrents. Because the water was unusually low they were able to creep along the jagged rock-walls safely and then come out on a bay beyond.

Mexia established camp a few miles above the Pongo at the mouth of the Rio Santiago, whose headwaters were in the Ecuadorean Andes. The group was in the dense forest of the upper Amazon along the first and eastern-most range of the Andes. She set up camp as best she could, glad when Jose joined her a few days later. They sent the canoes and paddlers back down the river and settled in.

Soon after they encamped, the rainy season started. Mexia later wrote, "The rainy season began with unprecedented violence and the rivers rose and rose until the roar of the Pongo could be heard for miles."

She, Jose, Valentino, and Neptali camped there for three months. The heavy rains made collecting difficult. Most of it had to be done from a small canoe because of the denseness of the forest. She was able to make short excursions, always by canoe

The Perfect Specimen

except for the day she climbed to the crest of the Sierra del Pongo. She also enjoyed the friendly Aguaruna Indians living close by. Jose knew a little of their language so they were able to barter trade goods including chickens, plantains, hearts of palm, and manioc roots.

Christimas 1931 was spent there. Mexia set up a little palm-tree under her thatched shelter, trimming it with wild red peppers and poinsettias. Then she hung some simple presents for the three cholos and some "mystified" but delighted Indians.

It soon became clear that the immense rains had them trapped. Mexia realized they had to be prepared to leave when or if the rains ceased.

The downpours paused in January and the floods temporarily subsided. They loaded the raft they'd made from balsa wood with Mexia's collections of plants, birds, and insects, the equipment that had survived the months of drenching, the four of them, and a tiny baby monkey Jose had acquired. Rafts were used extensively on the river system of the Upper Amazon. Because they were unwieldy, their course could only be roughly directed. A palm-leaf thatch hut was built over the platform of this raft. At the rear was a chicken coop to hold their remaining poultry. Valentino had built a fireplace, and he prepared their meals. They loosed the vine rope holding the raft and swung out of the Santiago into the Maranon.

Valentino and Neptali handled the big oars on either side. Then they were swept into the Pongo. The raft was caught by the racing current and tossed about like a straw. A whirlpool caught them, whirled them around three times, then "spewed us out." They sped on their way safely past another great whirlpool. In twenty minutes they had raced through the most dangerous part of the river. The gorge widened. They didn't have time to think about safety. They were rapidly carried into a circling backwash that swung them around and around in spite of Valentino's and Neptali's efforts. The rocks were jagged and they had a difficult time controlling the raft. Then "a lucky thrust" pushed them out into the current once more and they floated down the river at good speed.

Durlynn Anema, Ph.D.

As dusk approached each day, the men gradually worked the raft toward the shore. Sometimes they could find places to land, other times they would be swept on down the river. A curtain and a cot would be put up for Mexia, and the men would sleep on the raft's wooden floor.

As they approached Barranca, a boat came out to meet them with a huge packet of mail, some of which was nearly six months old. As Mexia read her letters and floated down the Maranon, she knew when they reached Iquitos both her raft journey and the trip up the Amazon would be finished. She would fly from Iquitos to Lima on her way home.

However, as she gazed at the river and thought of all she had seen and experienced, her heart -- despite the collection of 65,000 specimens -- was heavy as she left behind this once-in-a-lifetime adventure.

CHAPTER TWELVE

FINDING ECUADOR'S WAX PALM

"Ecuador, the land of the Equator! There my life-long shadow would dog my step no longer, but, vanquished, would grovel beneath my feet. At last I would stand on the earth's great belt, nearest the beneficent sun. But would it be beneficent, or would it strike me down with its invisible power or burn me with its intense rays?"

This is the opening paragraph of Mexia's Sierra Club article about her year in Ecuador -- "Camping on the Equator."

This collecting trip would be different from those in the past because she had funds for her collecting from the United States Department of Agriculture. The Department wanted specimens of plants that control soil erosion -- and Mexia was excited to accomplish her assignment.

'In addition to her responsibilities for the government, Mexia was determined to acquire a sample of the wax palm tree, or palma real. The wax palm Mexia sought grew on the Volcan de Chiles, one of the lower peaks between Ecuador and Columbia. She was interested because the wax palm was reported to grow at greater altitudes and to endure greater cold than any other known palm. If this were the case, it would adapt easily to California's varying locations and climate.

Mexia's fame as a botanical collector had spread. Consequently, she continued to be invited to speak. Now one of her favorite topics was her trip up the Amazon. The fact that she had lived in the jungle was as fascinating to listeners as her collecting talents. They listened in awe as she described her journey across the continent of South America at its widest part. Few women had accomplished feats of this type, so she was looked upon as a heroine as well as a mentor for younger women.

Both the Sierra Club and the Save-the-Redwoods League remained an important part of her life. She also tried to continue her studies at the University of California because she was, after more than a dozen years, close to a bachelor's degree.

As previously mentioned, she managed to visit with Dr. Brown and his wife whenever she was home. They also received letters from her wherever she happened to be. When Dr. Brown died, his wife had all of Mexia's letters retyped for posterity.

However, speaking and even studying took second place to Mexia's first love, exploring for new specimens. In 1935, she was sixty-five years old and far more active than people twenty years younger. She still could climb steep trails, though perhaps a little more slowly and taking deeper breaths. She was not afraid to camp with only her guides, nor to venture into regions where few women had gone. She was always eager to travel, especially in South America.

Her new adventure was called "Ecuador, Land of the Equator," because her other South American equatorial trips had been in warm climates. She expected the same weather in Ecuador. Then she started up the Andes and rapidly discovered a different equatorial climate. It was cold.

She left San Francisco on a steamer -- the *Santa Rosa* -- in September 1934 with her usual array of equipment -- presses, collecting materials, and camping gear. This ship was going through the Panama Canal, stopping on the west side at Balboa. Because she had to wait for her next steamer, she decided to ride the length of the canal, then return to Balboa via railroad. She had hoped to see some of the canal vegetation on her ride but "the train sped along so fast it was mostly a blur."

While she was in Balboa she took one excursion to Barro Colorado Island to view the animal and plant reserve. The remainder of the time she studied her materials, realizing she would have to send "interminable reports and accounts to the government."

While she enjoyed her ocean voyage and the canal trip, she was happy to arrive in Guayquil so she could begin her project.

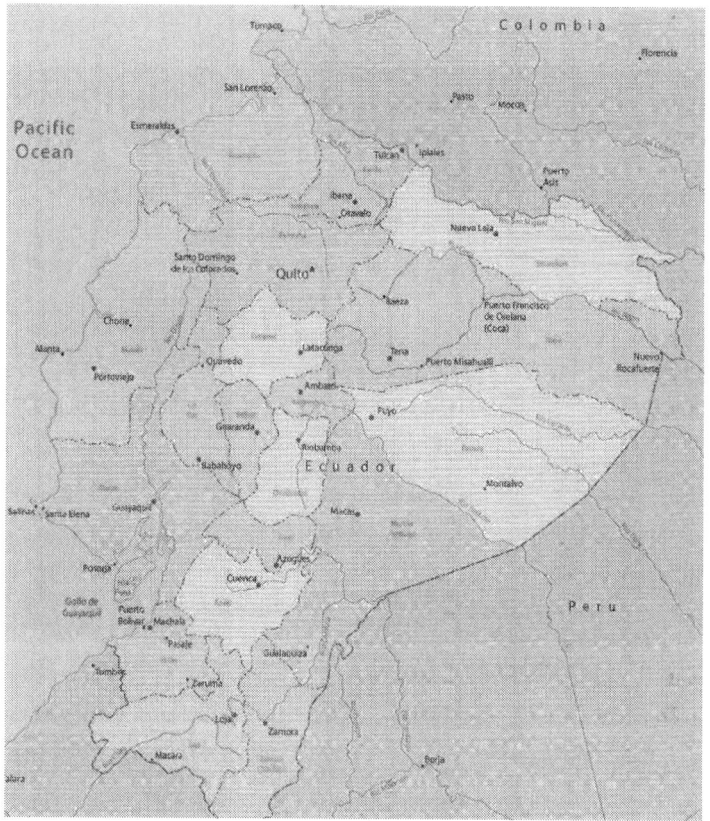

Because of the variations in elevation within Ecuador's boundaries, the climate ranges from tropical in the lowland region (La Costa) to cold and inhospitable in the highlands (La Sierra), despite being less than 150 miles apart.

Durlynn Anema, Ph.D.

In San Francisco, she had met a gentleman, Mr. Calderon, who had sailed earlier to Guayquil. He met the ship, helped her through customs, and escorted her to the hotel, which he helped operate. She was installed in the best room with three meals a day for $1.00.

Mexia now found herself somewhat of a celebrity because of Calderon, who told everyone a celebrated scientist was about to visit. Government officials greeted her at the ship.

She met the Governor and the American Consul as well as having schools, libraries, and museums ask her to speak. She met such requests dutifully because her "mountain" of baggage was not ready for the tri-weekly train. However, she was grateful when she finally could leave for her next "adventure."

On the train to Quito the climb up the Andes began. She noticed that the humidity abruptly diminished because the mountain slopes had little native growth after the lush tropics she just left. There were several switch-backs as the train climbed. When it stopped for the night before continuing northward, she saw majestic snow-crowned volcanos.

Mexia established residence in Quito for the year. Ecuador lies in three zones: the hot, low lying coastal plain with alternating wet and dry season; the mountainous region of high paramos and elevated peaks with scanty rainfall; and to the east the Oriente or Ecuadorian portion of the Amazon Basin. She planned for excursions in all three zones.

Ecuador presented the same kinds of challenges the sixty-five-year-old Mexia had been encountering and conquering since she began seriously pursuing botany fifteen years earlier. She and her team had to navigate the Andes mountains. The weather shifted dramatically in the high altitudes, often becoming quite frigid. As always, she had to adjust to a different culture, where time schedules weren't given the attention she thought they deserved, and endure the curiosities of some people who had never seen a white woman before.

Her first "adventure" during November was to the foothills of the Andes. She later wrote, "...if these are the foothills I am wondering what the real mountains are like! The plain comes up

The Perfect Specimen

to the hills and then these go dizzily up and down in sharp knife ridges like house roofs, so to speak, up and down which means the so-called trail pitches. The horses slither down one side in mud, tree roots, holes and slides and one hangs over their tails, then they clamber straight up over big rocks, more holes, big logs, roots, etc to the ridge."

She was staying in a Hacienda. When she presented her letter of introduction to the administrator of the several haciendas in the area, he tried to figure out why she was there. She attempted to explain, but realized he had never had a visitor like her. Her collecting trips were done on horseback because of the terrain. She rode a bay stallion; her guide Palma on another horse; helper Jose on a third a horse; and then "a rather small mule" carried all the equipment. She was especially taken with the tree ferns, finding one Cinchona or Quinine Bark. When her presses were full she returned by horseback to catch a launch in Santa Lucia to Guayquil.

The Oriente was her next destination in February. This eastern Andean province had only Quichua Indians. The only "whites" were Dominican Fathers at the Mission where Mexia was given a room while there. She said, "I am continually surrounded by a circle of the Indians quite agape at the strange individual and her belongings. Every article is picked up and examined, but as dishonesty is a vice of civilization, not the smallest thing is taken."

During her stay in this remote section she discovered a certain trait about all humans. "Yet the farther I stray and the more queer corners I visit the more I am convinced that after all life is pretty much the same everywhere. Here I eat (what there is), sleep on my cot and work collecting and caring for my plants, and the Indians go about their business of living and making a living and feeling the ordinary sentiments so familiar to us without great differences except in non-essentials."

She hiked this section, finding it very difficult. One day she hiked "up and down terrible trails" from 6:30 a.m. to 4:30 p.m., saying at the end "And was I footsore!"

Canelos was her main collecting point. According to Mexia, "It is a beautiful spot when finally reached, lying in a cup rimmed around with blue, forested mountains while a river, the Bobnassa, circles through it. The Indian dwellings are scattered, each far from its neighbors and imbedded in trees."

Here she hired an Indian guide, Juanucho, who was the "head Indian." She noted that "when in ceremonial attire he dons blue trousers, scarlet blouse, a bead necklace and has his face painted in intricate geometrical designs." They collected palms and the "cascarilla" or quinine bark tree. While they chopped down the palms, Juanucho would not allow her to do the same for the "cascarilla" because the Indians use it medicinally. She could only take the leaves and a bit of the bark.

Still in the Oriente she had an experience she would never forget -- riding not a horse who couldn't get through the "mire" of the trail, but an oxen. The huge white oxen which she called "El Palomo" had a sort of saddle on him. She was perched on top and she felt like "a midget." Four other oxen were packed. Two "ox-ateers" guided the animals along the trail at a pace of about two miles an hour.

". . .my Palomo was a treasure. Swamps have no terrors for bovines. When he came to one, Palomo would hesitate a moment, turn his head to several possible routes, select one, then plough serenely through the mire up to his knees, up to his belly, half way up his ribs, and sometimes he laid his chin on the muck."

All Mexia's collecting adventures were recorded in her vivid descriptive prose and sent to friends, then later used in articles.

Mexia's greatest Ecuadorian adventure involved the Wax Palm which she was determined to obtain.

Her determination meant she asked continually about wax palm locations. When she was back in the highlands she heard about one growing on the Volcan de Chiles, a lesser peak on the border between Ecuador and Columbia. Because the exact location was not told, she and Palma had to discover it. She, Palma, and Jose went by train to Ibarra where she obtained assistance on a possible location. Then an automobile took her to

The Perfect Specimen

Angel, a very small town (which wasn't very angelic she later said). Here she experienced the cold of the high Paramo. From there they went to a border town -- Tulcan -- where she had a letter to the customs officer.

He explained that the Volcan de Chiles was a snow-peak with the upper reaches consisting only of rocks covered with eternal snow and ice. This was above the tree line. If she wanted trees of her description she would have to circle the mountain to the north and on the eastern side drop down to the tree level. This seemed reasonable so she made plans for her trek.

Finally she found a packer who had two saddle and two pack animals. He claimed it was a terrible trail. "He was right!" she said later.

It was quite a trip. They had to spend a night in the cold where she took command and assembled a shelter of her ponchos. The next day the nonexistent trail meant the horses slipped, plunged, and went down in Alpine bogs along the steep hillsides. Then they made a steep descent. It grew rapidly warmer as they descended further and the forest thickened. They collected as they descended.

As they came to another village, they saw a Wax Palm but it was on an inaccessible cliff overhanging the river so they couldn't get it. Jose had a friend in a small "hamlet" who said he did know of one other wax palm but it was a half day's distance by a rough trail.

Mexia was excited. She met the friend and explained her reasons for seeing the wax palm. The next day he agreed to guide them. That trail went right up the side of a mountain so steep Mexia had to be hauled up. They went through choked forests and more steep slopes. Then they found it.

What a thrill for Mexia to be next to it, to see and touch what she had been searching. How she wanted to simply take a photograph but she knew she had to have it. "With a pang (unfelt by my companions) I gave the order. The ax bit in, and the great tree crashed to earth."

Mexia had obtained her wax palm, a major find.

Unfortunately, the journey ended on a down note. On the return trip, she was collecting and saw some blueberries.

Thinking they must be the same as in Alaska, she ate them eagerly. Immediately, she began to feel sick and dizzy. She wondered if it was the altitude. By the time her expedition reached the small-hut village of Tambo, she could barely stay astride her horse. Violent chills and pains shook her body.

The natives told her she had eaten poisonous berries. They dosed her with molasses-water and her guide, Palma, carried her to a cot. When the chills and pains increased, the natives captured a chicken and took a feather from it. Then they poked it down her throat. Soon she vomited the berries and everything else she had eaten. Quickly, the pains subsided.

Once Mexia recovered she pressed, dried and cared for her plants. She had heard there was another locality in Columbia with Wax Palms but there wasn't time to go there.

However she had reached her final goal in Ecuador -- finding the Wax Palm.

CHAPTER THIRTEEN

SOUTH AMERICAN ADVENTURES CONTINUE

"I am not at all certain of Mrs. Mexia's next move because she wasn't!"

Nina Bracelin wrote these words in a January 1936 letter to a Dr. Johnston. Mexia was in Santiago, Chile, after having been in Peru during the Fall. She had not decided what she would do next.

Bracelin asked Dr. Johnston to write to Mexia about his ideas for collecting in Chile. "She seemed a little uncertain as to ideas for localities, hence my suggestion to you. I can not imagine her staying in Santiago for any length of time."

After the U. S. Department of Agriculture sent her to Ecuador, Mexia wanted to continue her travels in South America to see as many places as possible in the next year. She felt, after making such a long trip to Ecuador, that she might as well take advantage of it and see more of South America.

On October 1, 1935, following her year in Ecuador, Mexia joined Dr. Thomas H. Goodspeed and his wife in Lima, Peru, to collect in that region. He was a botanist from the University of California, Berkeley. He and his wife were a couple with whom she could work. Eagerly, she met them to collect south and west of Lima. When the Goodspeeds decided to go to the southeast, Mexia left them to go north to Huanuco in the Cerro de Pasco region to collect on her own. She agreed to meet the Goodspeeds

Durlynn Anema, Ph.D.

and others at Mendoza, Argentina at a future date. Her original plan had been to stay with the Goodspeeds for the duration of their trip, but collecting on her own was so important to Mexia she had to depart.

Among the specimens she searched for were the nicotianas. These tender shrubby plants are native to the Americas, with the tabacum, a type of nicotiana, as the commercial source of tobacco. The longiflora variety has white-to-purplish flowers and is found from Peru to Argentina. In her quest she went from the highlands of Peru to several destinations in southern South America.

The search became very broad. Mexia took a steamer across Lake Titicaca, marveling at the beauty of the highest navigable lake in South America. Next came a trip by train to La Paz, Bolivia, where she not only enjoyed the high altitude town and its occupants, but was able to do more collecting.

Another train took her to Tucuman and Mendoza in the Argentine Andes, although she didn't meet the Goodspeeds there. Then she traveled by car via Puente de Inca, a natural bridge, to the crest of the Andes. She had accomplished a sightseeing goal, to view the Christ of the Andes, a massive statue sitting atop the border between Argentina and Chile--the area known as the Uspallata Pass. The governments of both countries commissioned the statue in 1904 as a symbol of a peace treaty they had signed. The statue moved her deeply so she spent a great deal of time admiring it and the surrounding scenery.

With her arrival in Chile, Mexia realized she was close enough to fulfill another dream-- traveling to the farthest point south on the continent. She took a train to Puerto Montt, Chile, where she boarded the *Alejandro*. It was a crowded steamer going through the Chilian Fjords and the Strait of Magellan to Punta Arenas, Chile.

THE PERFECT SPECIMEN

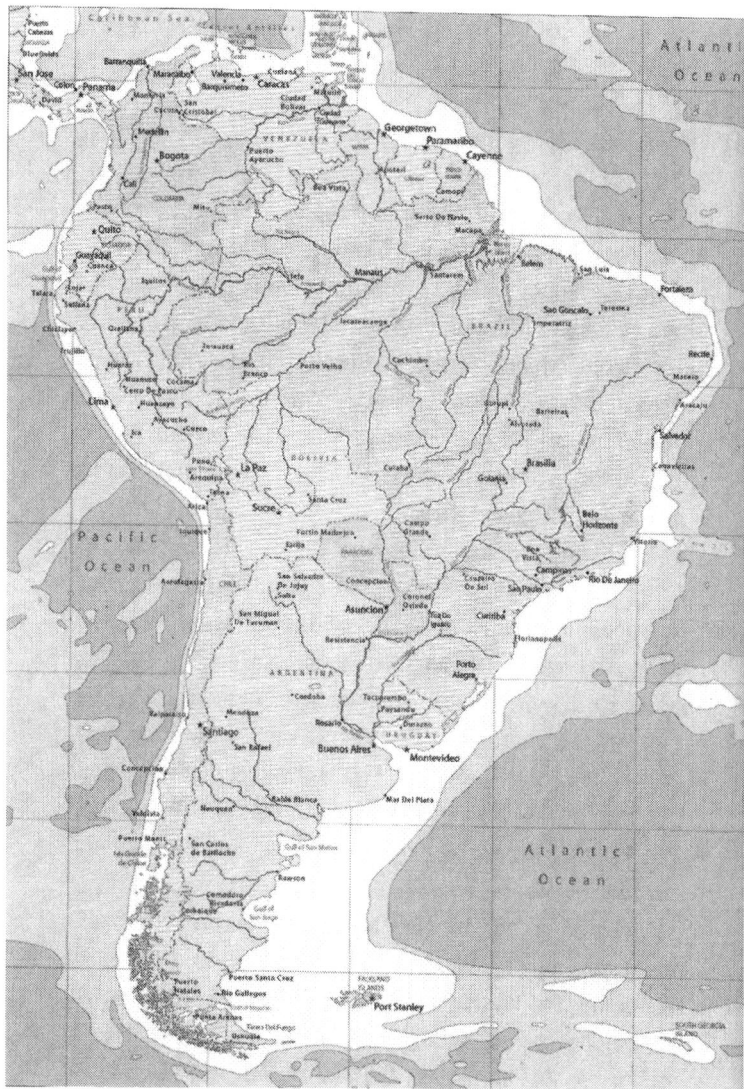

Mexia traveled throughout South America after exploration in Brazil and Ecuador. She traveled through Peru, Argentina, and Chile to Tierra del Fuego in the South.

The Chilean Fjords are some of the most spectacular on Earth. Mexia regretted not landing to explore. *Photo Courtesy of Durlynn Anema*

The Perfect Specimen

During the long trip, the steamer stopped only at the island of Chiloe, in southern Chile, which was a great disappointment to Mexia. She wanted to visit and explore as many places as possible, and felt she had been denied this privilege. However, the fascinating scenery through the Chilian Fjords and the Strait of Magellan kept her at the ship's rail, although she was disappointed that some of the trip through the Strait was during the night. She later wrote that this inland passage was "still more beautiful than the passage to Alaska. The innumerable islands, many of them still unexplored, are but the peaks of a submerged mountain chain and are forested to the water's edge with Nothofagus, the evergreen Southern Beech. Beneath the trees they are said to be waist-deep in spongy lichens, but I had no opportunity to land to verify this."

At one point, Alcalufes Indians paddled toward the steamer in large canoes. They wanted to barter nutria and fox skins for food, but the *Alejandro* did not stop, much to Mexia's disappointment. According to Mexia, "These were the original 'Canoe Indians,' that now had dwindled to a rare few. They lived mostly in their bark canoes and carried always on a heap of sand glowing firebrands. It was due to these flickering fires seen among the countless islands and not to volcanic action that Tierra del Fuego acquired its name. They lived between the islands and along the Strait, with their diet mainly shell fish."

Letters from friends helped Mexia obtain passage to Rio Grande on the eastern coast of Tierra del Fuego. There she stayed with the manager of a big frigorificos (slaughterhouse). She was surprised at the many people inhabiting this area, which she thought would be an icy, uninhabited wasteland. To her eye, most inhabitants were English, Scotch, Scandinavian, or Czech. Huge estancias (sheep stations) dominated the region. In season, up to 5,000 or more lambs were killed each day and sent in refrigerated ships to England. "This, being an oceanic, not a continental climate, has not these extremes. They have some snow and ice in winter, but I am told it does not last long, and sheep, cows and horses winter without shelter."

While Mexia enjoyed visitng Tierra del Fuego, she was disappointed the time of year did not produce more exploration. *Photo Courtesy of Durlynn Anema*

Unfortunately, the sheep had devoured almost all the vegetation. Further, Mexia was there at the end of the season and the frost had already set in, withering the flowering plants. She did enjoy the Southern Beech whose leaves had turned burnt orange, russet, and vivid crimson due to the arrival of autumn. She was sorry to hear the beeches were being steadily cleared off to provide more pasturage for sheep. The family with whom she was staying took her inland to find more plants -- but her collecting was not as abundant as other places.

Because the autumn rains were about to arrive, Mexia decided to leave the mountainous area and return to Punta Arenas. There she boarded the *Alejandro* for the return trip to Chile.

THE PERFECT SPECIMEN

On Mexia's return trip to Chile, she was able to stop at Chacabuca, Chile. *Photo Courtesy of Durlynn Anema*

Mexia was also able to stop at. Ponte Mott, Southern Chile, she enjoyed it for its beauty and abundance of collecting.
Photo Courtesy of Durlynn Anema.

THE PERFECT SPECIMEN

In late May 1936, she was back in Lima, Peru. She wrote, "After all Peru is perhaps the most interesting of the nations bordering the Pacific, for besides the three grand topographical regions of low, hot coast; high Sierra and Plateau; and densely forested eastern tropics, it has the remains of the vanished Indian Empires of Incas and Pre-Incas."

A steamer took her to Mollendo, Peru, which she called "a forlorn little port." From there she took a train to the mountains, eagerly watching the changing scenery. "The railroad is a marvelous one. For a number of miles it runs along the beach, with the surf breaking a few feet below, then turns inland and begins to climb. As it practically never rains on this Humboldt-current-bathed coast, the hills are absolutely bare. As one rises, an occasional spindling cactus shows itself, and when the track reaches a trickle of water in a ravine, it follows this until a fair sized stream plunges down the canon (sic) along which the road winds. The mountains grow steeper and more rugged, and snow peaks peep over the shoulders of the nearer ridges, while the train thunders around countless curves," she wrote to friends.

Mexia first stopped at Cuzco, then took a train over the pass and into the deep gorge of the Urubamba River. From there she took a bus to the town of Quillabamba. This town was several hours' ride down a narrow valley along the railroad bed. She first tried collecting near town, but cultivated fields of corn, cocoa, and coffee were the only vegetation. She and her guides then rode horses up the river to find vegetation on the hills. Here she found and filled her press with "a beautiful rose-red bougainvillea which wreathed the streamside trees." Beside collecting, she enjoyed talking with a Dominican priest who had spent thirty years among the Machiguela Indians and had many tales of his experiences.

When the opportunity arose to visit Machu Picchu, she eagerly went. As Mexia approached the ancient ruins, discovered in 1912, she could hardly wait for her first view. Her horse was slow, but finally she spied the buildings, "roofless but otherwise as perfect as the day when, time and reason unknown, the inhabitants last drifted away to leave their city on a hill top to the silences of the dead, to the engulfing vegetation and to the owls and snakes."

Only a day and a half was spent at this fascinating ruin. Mexia explored until dusk the first day and was up at dawn the next day. "It took the sun a long time to light up the eastern rim, then it shone on the mountains to the west and traveled slowly down their precipitous flanks. I watched it all, entranced, no life but my own to see it where before, in days long past multitudes chanted a welcome." She tried to see everything and wished she could have stayed longer. But her bus back to Quillabamba was waiting.

Mexia stayed in Lima for several months, taking short collecting trips to various parts of the country, all of which she found interesting. Rich collecting was found in several of these terrains.

> In a letter to Alice Eastwood, August 18, 1936 she wrote: "The climate, scenery, soil, altitude, vegetation, etc. are so exceedingly varied that each portion of the continent is a country to itself. The coast is rainless desert practically without vegetation of any kind except where a river comes down. The high Puno is rocky, cold, and with only scant Alpine plants which diminish up to the snow line. In between there are many valleys where different types of vegetation grow. Then again you get the bare, arid mountains that surround chilly Lake Titicaca or you drop down to the lush, eternally green forest of the eastern slopes of the Andes."

On November 29, Mexia boarded the ship *Manezeles* for the trip to Esmeraldas, Ecuador. She was looking forward to yet another adventure. Arriving on November 23, she stayed to write an article, then sent it and ninety-five Christmas cards home to San Francisco. She heard a few days would elapse before a boat

THE PERFECT SPECIMEN

could take her to Limones, Ecuador, so she unpacked cases and boxes to repack what she would need for an inland trip. She was busy sorting out her things when she was told the boat had arrived and would sail at dawn the next day.

When they reached Limones, Mexia left the boat to buy some supplies. A young boy helped carry her purchases as she walked back to the boat on a plank. When she began across the hatch, she fell through a broken board, severely injuring her leg. While the skin was not broken her thigh took the weight of the fall and turned black, making walking almost impossible.

As was her custom, she persevered, continuing on the boat from Limones to Concepcion. A canoe took her upriver to Playa Rica, then to her camp at El Sajedo. She stayed there from December 5 to December 28, did what collecting she could while pressing and caring for some of the thousands of plants she had collected in the preceding years, including 13,000 in Peru alone.

Her reverse trip started on December 28 and culminated on January 8, 1937, when she caught a ship for the Panama Canal and San Francisco. She arrived in San Francisco on February 5. Mexia was happy to be home -- if you could even consider her to have a home at this point -- so she could sort out her collections, give directions to Bracelin, and continue her studies.

CHAPTER FOURTEEN

FINAL EXPLORATION IN MEXICO

"Now that I'm back after more than two years in the wilds of South America, I find myself longing for a nice quiet jungle again," Mexia told a *San Francisco News* reporter on March 6, 1937.

Mexia was sixty-seven years old and still anxious to visit as many places in Mexico and South America as possible. She also wanted to finish the one course she needed for a bachelor's degree.

Bracelin collected and saved all the letters Mexia had written, fully understanding this woman was now a renowned collector. She not only preserved those written to her personally but also ones to other correspondents.

Mexia had written a great deal about all her adventures. She hoped to sell one article -- "Following the Sun across South America or Up the Amazon and over the Andes"-- to *National Geographic*, but her dream never materialized. However, parts of the article found their way into the *Sierra Club Bulletin* and *The Gull*. Mexia also published in *Bird Lore, Better Health*, and *Madrono*.

Mexia, always eager for new experiences, never stopped learning. When she heard, in June 1937, about the biological station at the University of Michigan Ann Arbor, she applied as a student and was quickly admitted. She spent the summer at the station in Cheboygan, Michigan. However, she became more a teacher than a student. She happily presented her extensive botanical knowledge to the students, yet felt she gained even more knowledge during her stay.

After her Michigan studies, she traveled to Philadelphia to visit her sister Adele, spending the end of August and part of September there. Adele, who never married, had remained in the eastern United States. Mexia hadn't seen her since May 1934, following a university lecture she had given in New York City. They maintained a correspondence throughout the years, but opportunities to see each other were rare.

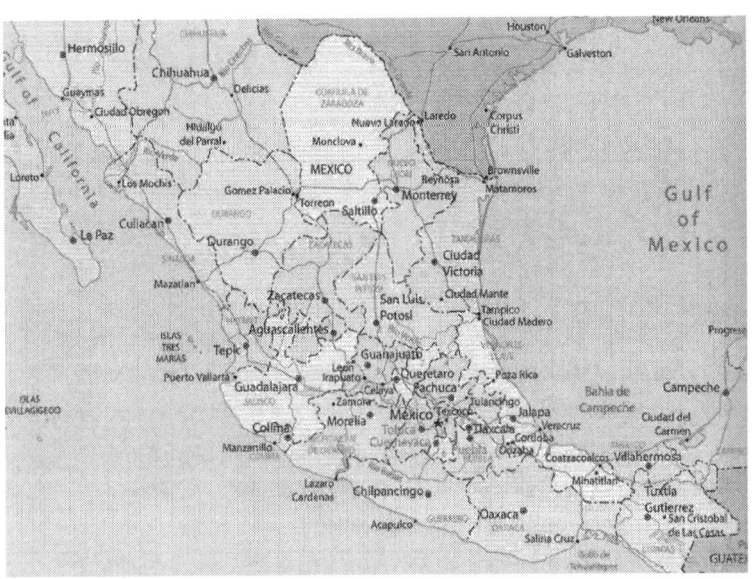

During Mexia's 1937 Mexico trip, she visited several small towns in the country's southern inland region near Mexico City.

The Perfect Specimen

The wilds still beckoned Mexia. In October 1937, she started another trip, this time to areas around Mexico City and Oaxaca. Her first week was spent going to a mine in the State of Guerrero. Trekking the steep slopes, she hoped to find more than she did. Disappointed, she decided to return to Mexico City. Also, while she didn't want to admit it, she was not feeling well and spent the next week seeing doctors and trying to recuperate. Feeling weak and tired bothered Mexia, but she was determined to overcome all signs of illness. During her time in Mexico City, she took day trips to Toluca, Cuernavaca, Taxco, and Puebla.

By October 26, she was feeling well enough to leave Mexico City and travel to Balsas. She traveled by canoe on the Balsas River to a mine in the state of Guerrero. The mine was quite near the coastal belt and very hot. The hills were turning brown because this was the dry season, and again she found little to collect.

Mexia hired a mozo, Severo, to go further into the countryside. She was pleased with her choice. He knew the country and was reliable and helpful. "The old man, Severo, has turned out pretty good, old and somewhat slow but reliable and willing and tries to follow every little indication I give him." They collected vegetation along the streams.

She also hired a second helper -- a school girl of 15 who she called "quick and smart but flighty as youngsters are and wants to slap quickly through things."

In the same letter Mexia mentioned that it continued to be hot and dry. "I stand the heat pretty well, but it was 90 degrees in the shade this afternoon and I am much of the time in the sun and walking." The house where she stayed was at the top of a steep hill, whose climb became more difficult for her each day.

One day, as she was crossing a little river and jumping from rock to rock, her boot slipped and she went knee deep into the water. She wrenched her right shoulder badly enough that she went to see the company doctor at the local mine. He said it was a severe sprain with no broken bones. His advice was hot applications, liniment and taking time to rest. Each day she could raise her arm a little more, but "probably it will be complicated by the rheumatism in my case," she wrote to Bracelin.

Her one other comment about the location was "personally I do not care so much for the flora of the half dry countries, but some people like it, I suppose, and this country has been little collected, I hear...the place is healthy and insects are not bad."

In the middle of December, Mexia arranged to go to a mine higher into the Sierra Madre. She and Severo went four days south of the mine with horses and pack mules. As they left the river basin the mountains became higher and steeper. At first, the mountains were covered with deciduous oak trees and jungle, which merged into forests of pine. At 5,400 feet, Mexia and Severo reached the last habitation. The native farmers lived in primitive conditions, yet still offered what food they had -- beans and corn tortillas.

Mexia wrote to Alice Eastwood, "While I lost weight 1 did not starve... the accommodations were poor (but) the country was gorgeous. There was wave after wave of the Sierra Madre range all covered with the heavy, untouched pine forest and trees well up to two hundred feet high. The oaks still were present, also madronos and some other few broadleaf trees."

She continued in the Eastwood letter that "the collecting was fairly good for the dry season, and daily Severo and I would work along one of the arroyos. I had a sorrel horse that was fine on the levels or uphill, but going down he would plunge and want to run. In one of his plunges he fell on a steep slope and went down as I did also, but luckily free of him. However, I lit on my thigh and was lame for some days and black and blue from hip to knee. It has all cleared up now."

The collecting here also was scanty. However, Mexia was enthralled by the adiantum. This is a plant commonly known as the maidenhair fern, of which she never had seen so much in her life.

Christmas was spent as any other day -- collecting. Mexia wanted to go a day's journey further to Las Lumbreras, which was the site of a lost mine. However, she could find no one to take her until New Year's Day. Then a shepherd and his family,

THE PERFECT SPECIMEN

who had gone for supplies and now were returning to Las Lumbreras, agreed to guide Mexia and Severo. They packed a burro, saddled their horses, and started out.

The shepherd and his family were hospitable, sharing their food and campsite. The shepherds in that area ran mostly goats and a few sheep. They had no house, sleeping in the open. At night, the entire family -- father, mother and three daughters, all covered with serapes -- slept right next to each other to keep warm. They invited Mexia to join them but she declined.

When she arrived at Las Lumbreras, she discovered the sheep, goats, and cattle had grazed the grass, shrubs, and low trees to the point where collecting was difficult. After a day, she and Severo decided to start back. They had been gone for five weeks, and Severo wanted to return to his family. On top of that, Mexia was beginning to feel ill again. They found mules for their packs and returned to the mine, collecting along the way. Mexia returned to Mexico City.

Although still not feeling well, she set out in March to collect in the Oaxaca area, south of Mexico City. Before she left, she went to Mexico's Department of Forestry, Fish, and Game. To her surprise, she found an old friend from her early days in Mexico -- Juan Zinzer. He was thrilled to see her and took her to his superior, who gave orders she should receive all possible assistance. Then he made her an honorary member of the department.

Mexia mentioned in her letter to Eastwood that "I have been away from Mexico, my home for many years, so long that I felt myself among strangers, but when I went to the Department of "Forestal, Pucu y Caza (Department of Forestry, Fish and Game) I found Mr. Juan Zinzer."

When she arrived in Oaxaca, she surprised her hosts by going to their home unannounced. She had her mozo Daniel, one saddle mule, three pack mules and three arrieros (or packers) with her. Her surprise visit was to Don Guillermo-Barth and family who she had known when she lived in Mexico years previously. He had been a coffee grower for many years, had a German wife and a daughter who also was married to a German.

Mexia said "They are very hospitable ... and so I am well content."

Once again she was ready to collect with her mozo Daniel. They gathered pack animals and set out for the higher mountains. However, it was raining and did not stop throughout the night, plus it was very cold. They climbed even higher to be enveloped by a chilly mist and drizzle. While she saw much she would have liked to collect the weather was too miserable to proceed "Had to be reluctantly left untouched."

Mexia stayed out on this trip for a week, camping as usual. She looked forward to her cot the first night but found it less than cozy with a biting north wind going right through her covers. She then put on all the clothing she had, then added her warmest clothing and the rubber poncho on top. This kept her warm for the remainder of the trip.

While she found plants, she was appalled by the grass fires set by the natives in order to plant crops. The trail followed knife ridges and skirted mountain flanks, or dipped down to streams still in mist during the morning. In the afternoons, she and Daniel came into sunshine to see "glorious" pine forests as they warmed up. Their last three evenings were in grassy, well water locations "where the tired mules luxuriated." In spite of some discomforts, she did enjoy the "fine oaks and splendid pines covering the slopes and crests."

At the end of April, Mexia's weakness had returned. She saw a doctor who said her condition was more serious than she anticipated, and recommended her returning to California and her own physician. Reluctantly, she took his advice.

Before Mexia left, she took great pains to arrange all her equipment and put everything in order, expecting to return soon.

Chapter Fifteen

Leaving a Broad Legacy

Bracelin and Dr. Brown met Mexia at the dock in San Francisco on May 29, 1938. Brown immediately put her in the hospital, where she was diagnosed with lung cancer. Mexia accepted the diagnosis as she had accepted so many other things throughout her lifetime -- this acceptance that she may not have long to live. Bracelin was devastated that her friend and mentor now would be with her for only a short time. While they had kept in touch by mail, they seldom had been together physically because of Mexia's lengthy adventures. Now the time was even shorter.

Bracelin later wrote to Dr. A. C. Smith, a botanist for the New York Botanical Garden: "Fortunately, she had come home as a physician in Mexico City directed her to do, and she was here where she was given every care within the power of medical science as well as being with the two persons who meant most to her, her physician and myself. I spent most of my time with her and was alone with her at the end."

On July 1, Dr. Brown advised Bracelin it would be best for her to take Mexia to her home. The hospital atmosphere was only making Mexia depressed. He hoped the more pleasant atmosphere and better food would revive her both spiritually and physically. Bracelin was only too happy to comply.

Durlynn Anema, Ph.D.

Mexia was taken to her Berkeley home, where Bracelin lovingly cared for her. Once again they could talk and laugh as Mexia reminsced about her adventures and her collecting. Mexia enjoyed the quiet of the garden and the deep friendship of a woman she considered colleague, daughter, and closest friend.

Despite all efforts, Mexia died on July 12.

A stunned Bracelin wrote to Dr. Smith that Mexia was "like an older sister or mother to me and I shall miss her greatly."

Mexia's death was mourned by those closest to her and her interests -- the botanical community, the Sierra Club and the Save-the-Redwoods League. William E. Colby, Councilor for the Save-the-Redwoods League, became executor of her last will and testament and also wrote Memorials about her. Nina Floy Bracelin also performed that duty.

In Mexia's "Memorial" in the *Sierra Club Bulletin* Colby gave her history, then said: "This brave explorer and collector of rare and unknown plants was a warm friend of many of the members of the Sierra Club. Before she started her botanical exploration she went on many Sierra Club outings. These doubtless served to stimulate her interest in botany and on these she gained experience in living in the out-of-doors, a knowledge which later served her well on her many arduous trips of exploration and scientific collecting...All who knew Ynes Mexia could not fail to be impressed by her friendly unassuming spirit and by that rare courage which enabled her to travel, much of the time alone, in lands where few would dare to follow"

The Sierra Club also passed a resolution saluting Mexia: "Resolved that the Sierra Club, in the death of Ynes Mexia, has sustained a great loss...Because of her scientific knowledge and painstaking notes, the specimens collected by her have been recognized as of exceptional value. Mrs. Mexia became a member of the Sierra Club in 1917 and went on many of its outings and, because of her cheerful disposition and extensive botanical knowledge, added materially to these expeditions."

Bracelin wrote for *Science Magazine* a recap of Mexia's career as part of their Memorials section. Among her comments was: "... She collected approximately 9,300 numbers, from

THE PERFECT SPECIMEN

140,000 to 150,000 specimens and over 500 new species, the last collection being yet unidentified. Many new species and one new genus were named in her honor."

Mexia had prepared her estate carefully, wanting her legacy to be one of environmental protection. She wrote her Will on October 2, 1937 and carefully included everyone important to her, as well as the two organizations she so enjoyed as she regained her health -- Sierra Club and Save-the-Redwoods League.

First, she revoked all prior wills made by her.

Then she left money to a half-niece in Texas, to a Mrs. Ellen L.Moffett in Oakland, California, and to Dr. Phillip King Brown "to be expended by him in the manner known to him and me, but without any further limitation thereon..."

The California Academy of Sciences was given $3,000 "to be used by it in paying therefrom until said amount is exhausted the sum of One Hundred Dollars ($100) per month to Mrs. Myram Floy Bracelin so long as she shall do or perform for it botanical or other work..."

Then she bequeathed to William E. Colby of Berkeley, California, in trust for the period hereinafter stated for the following uses and purposes, that during the joint lives of her sisters, Amanda and Adele, he would hold the income from the money in her accounts and pay them equally. Upon the death of either sister, her remaining money would be divided equally between the Sierra Club and the Save-the-Redwoods League.

The Will was filed in the City and County of San Francisco on July 15, 1938.

Bracelin continued to work on Mexia's collections, which were incomplete at her death. She noted this in a letter to Ivan Johnston August 6, 1938. "I have all of Mrs. Mexia's last Mexican collection untouched and I wonder if you would not like to go over it when you are here? I am certain it would help me a lot but more than that I think it might be useful to you."

Throughout her lifetime, Bracelin wrote articles about Mexia and worked closely with institutions across the world to ensure Mexia's collections were intact. Because she had sent Mexia's specimens throughout the world since the time she started to work with Mexia, she had built up a wide correspondence and acquaintanceship with international botanists.

Her work became an important contribution to the record of horticulture in central California. She felt that while she had accomplished it, indirectly it was a bequest from the estate of Ynes Mexia.

Before her death, Bracelin deposited all the records of the Mexia collections, much of the Mexia correspondence and information about the Mexia family in the Bancroft Library, University of California, Berkeley.

Money held in trust for her family, upon their deaths, ultimately went to the Sierra Club and the Save-the-Redwood League. Her heart always remained close to these environmental organizations. The Save-the-Redwoods League received over $25,000 and used an additional $25,000 to purchase the Ynes Mexia Memorial Grove in Mendocino County, California. The Sierra Club received $20,000.

The Save-the-Redwoods League had a difficult time deciding where to use the Mexia legacy. During the 1950's, they discussed buying a grove in Prairie Creek Redwoods State Park in tribute to Mexia. This decision had to go before the California State Park Commission.

The League's final decision changed locations of the grove. On January 20, 1968 Bracelin received a letter from the Commission that they had received a gift from the Save-the-Redwoods League to establish the Ynes Mexia Memorial Grove at Montgomery Woods State Reserve in Mendocino County.

"We understand that Ynes Mexia had a deep feeling for the Redwoods and had a special interest in Montgomery Woods State Reserve. Those of us who know the quality of the Redwoods in this Reserve will share with you the satisfaction of the establishment of a Memorial Grove in your sister's name in this area."

The Save-the-Redwoods League then received a letter dated January 25, 1968 telling them "It is my pleasure to inform you, as requested, that the State Park and Recreation Commission at its meeting on January 12, 1968 in Pacific Grove, approved the establishment of the following groves in the specified units in the

The Perfect Specimen

State Park System:"... (five groves in Prairie Creek Redwoods State Park) then "Montgomery Wood State Reserve, Ynes Mexia Memorial Grove."

How pleased Mexia would be to visit her grove in Montgomery Woods -- a quiet, off-the-beaten path State Reserve that reflects her love for these giant and majestic trees. These are old growth Redwoods, like the ones she wanted to preserve. And flowing through her grove is Mexia Creek -- an extra tribute.

Mexia's legacy lives on in the new genus -- the *Mexianthus Mexicana* -- and fifty new species named after her, as well as the vast collections she obtained for university and museum herbariums across the world.

Ynes Mexia, the shy, lonely girl, became a worldwide name in botanical circles. Her dreams of exploration and collecting were just beginning at an age when many people retire or slow down. Age was never a factor for Mexia. She was too busy learning, leading the adventurous life few young people experience -- and enjoying every minute of her new found life after fifty.

Durlynn Anema, Ph.D.

Mexia Grove in Montgomery Woods State Reserve, Mendocino County, California. *Photo Courtesy of Evan Johnson, Save the Redwoods League.*

The Perfect Specimen

Mexia Creek in Montgomery Woods State Reserve, Mendocino County, California. *Photo Courtesy of Evan Johnson, Save the Redwoods League.*

Timeline

1870 Born in Washington, D. C., on May 24.

1871 Moves to Mexia, Texas, with parents and sister Adele.

1879 Parents separate; moves to Philadelphia, Pennsylvania, with mother and sister; attends schools in several cities in Eastern United States.

1880s Moves to Mexico City to be with father.

1898 Marries Herman Laue; lives with him in Tacubaya, near Mexico City.

1904 Herman Laue dies.

1908 Marries Augustin de Reygadas; lives at ranch in Tacubaya.

1909 Has mental breakdown and moves to San Francisco.

1917 Joins Sierra Club and begins to go on outings.

1921 Enrolls as special student at the University of California, Berkeley.

1922 Joins an University of California Berkeley botanical expedition to Mexico.

1924 Regains American citizenship.

1938 On May 24 returns to California from Mexico; dies on July 12.

MEXIA'S EXPEDITIONS

September-November 1925
Western Mexico (Sinaloa): 3,500 specimens

August 1926-April 1927
Western Mexico (Sinaloa, Nayarit, Jalisco): at elevations up to 6,000 feet in the Sierra Madre; 33,000 specimens.

June-September 1928
Alaska (Mt. McKinley National Park -- now Denali National Park and Preserve): first general collection of the park flora; 6,100 specimens.

May-July 1929
Northern and Central Mexico (Chihuahua, Mexico, Puebla, Hidalgo): 5,000 specimens.

October 1929-March 1932
South America (Brazil -- Rio de Janeiro, Vicosa, Diamantina, the state of Minas Geraes, the Amazon and other river tributaries in the states of Para and Amazonas; Peru -- upper Amazon and Santiago river valleys): 65,000 specimens.

September 1934-January 1937
Two and a half years exploring South America. Two major trips.
September 1934-September 1935
Ecuador -- coastal plains and eastern Amazonian slope of Andes, northern highlands and Columbia border: 5,000 specimens.
October 1935-January 1937
Peru expedition with Goodspeeds; Chile-- southern Chile, Strait of Magellan, Tierra del Fuego; Peru -- Cuzco, Machu Picchu, and Cerro del Pasco; Argentina -- Tucuman and Mendoza; Ecuador -- Esmeraldas: 13,000 specimens.

Durlynn Anema, Ph.D.

October 1937-May 1938
Southwestern Mexico (Oaxaca and Guerrero): 13,000 specimens.

Additional Chapter Information

Chapter One -- A Lonely Life

General Jose Antonio Mexia
Ynes Mexia's perseverance and determination came from her grandfather who was active in Mexican affairs until his execution in 1839.

Jose Antonio Mexia was born December 31, 1800 to Pedro Mexia and Juana Josefa Hernandez. According to Jose Antonio, he was born in Jalapa, Veracruz, but some historical records have his birth in Cuba. He lost his father and brother during the Mexican War of Independence and had to seek refuge in the United States where he became proficient in English. In November 1822, because of his proficiency in English, the Texas governor (at this time Mexico owned Texas) named him interpreter for a Cherokee Indian delegation to Mexico City. From that point, he became active in the Mexican government, Mexican politics and Masonry. He first served briefly in 1823 in the Mexican army as a captain. In 1827, he again entered active service and quickly received several promotions: lieutenant colonel in 1828; colonel in 1829: brigadier general in 1832.

Mexia served in the U. S. as secretary of the Mexican legation 1829-1831 and became an agent and lobbyist for the Galveston Bay and Texas Land Company. He was a supporter of Antonio Lopez de Santa Anna and in 1832 led an expedition (called Mexia's Expedition) to suppress a possible rebellion by the new American settlers. Stephen Austin convinced him the settlers were loyal to Mexico. Then, while a senator, he joined Federalist forces that rose against Santa Anna's assumption of dictatorial powers in 1834. After a two month campaign he surrendered and was ordered into exile by Santa Anna. He resided in New Orleans, continuing to try to overthrow Santa Anna. In January 1839, he returned to Tampico and joined forces with General Jose de Urrea. However, these forces were undermanned and on May 3, 1839 Mexia was captured and executed by firing squad.

On August 5, 1823 he married Charlotte Walker, who was twenty-two years old and English born. Two of the children -- Maria and Enrique -- were closely associated with the development of northeastern Texas during the last half of the nineteenth century.

Information from: Handbook of Texas Online, Raymond Estep, "Mexia, Jose Antonio," accessed July 7, 2018, https://tshaonline.org/handbook/online/articles/fme34

Chapter Two -- Adjusting to a New Life

Mexia letter to de Reygados explaining her reasons for separation.

"Dear Petsito;"

"Your letter of July 19th has reached me, and I think you have taken a very wrong and mistaken idea of what I wrote to you about coming. If I wrote telling you what your life was likely to be in this country, I did this through a sense of justice, because it did not seem fair to me to let you come without some idea of what your circumstances would be. ...

"In the first place you will have to work and make your own living, and you will have little or much according to what you earn. I do not think there is anything in this to hurt your feelings.

"In the second place I told you not to bring any plants or animals, not because I object to your having them, or forbid it, but because you will have to live in a boarding house, and your own common sense should tell you that it is impossible for you to have chickens or any other animals in a boarding house. ...

"The third point is my being unable to live with you as your wife, and I do not impose this on you as a condition on which I will allow you to come here, as you seem to think, for if you come here this will have to be exactly the same. It is not that I am displeased with you or am trying to punish you for anything you have done, on the contrary, I am very sorry that my inability to bear with the marriage relation causes you any trouble or discomfort or distress of mind. I only wish that this were not so, but see no way to remedy the situation. As you know I was very ill when we married, seriously so, though you are not willing to acknowledge it, and you know that I grew steadily worse after we were married until we separated. You know the great nervous disturbances which I suffered from the marriage relation, and how they exaggerated my condition, and yet you had no consideration for me.

"I endured this as long as I could, growing steadily worse, until I could stand it no longer, and we separated. Since then I have steadily improved ...

"I have asked of you no sacrifices that I have not made myself, and except this separation from me, I think you cannot complain; you have been in the country you prefer, have had your family and your sister to live with you, have had your animals and other things to your liking, and have been provided for. ... the only thing to do is to decide that understanding thoroughly the fact that I cannot live with you as your wife, how and where you will be happier...

". ... As I cannot live with you there is no necessity or duty for obligation for you to come, and if you come it will be because your own wish brings you here.

"Speaking of affection you knew before we were married (because I told you so) and afterwards yourself remarked it that I have never known any sex love, do not think I am capable of it, and the affection I have had has been that I might have had towards a brother. I will not deny that my fear and horror of the marriage relation has greatly diminished this, and I acknowledge the truth of your statement that I am not of a very affectionate disposition, or at all sentimentally inclined. I think it is more honest and fair to you to say this clearly, but what I can also say is that I take a very real interest in you and your happiness and well being. ...

(She signed it) Yours affectionately,
Pesita

Durlynn Anema, Ph.D.

Letter from Ynes Mexia Collection, Bancroft Library, University of California Berkeley.

Chapter Three -- Finding a Niche

Mexia letter to Mr. Robert G. Sproul, Secretary, University of California, October 11, 1920, regarding the redwoods in Montgomery Grove.

Dear Mr. Sproul:

I joined the League sometime ago, and have been very much interested in the movement for which it stands; I have given it a great deal of thought, and have been wondering if there is anything I can do, as an individual, to help the work along.

I have been much distressed to hear that this past summer cutting has been going on in the Montgomery Grove, which I visited last year, and -- I do not vouch for the truth of the story -- that the "Big Tree" there in that grove was only saved because a division line of two owners ran through it, and as one owner declined to allow his half to be cut, the other one perforce left his. When I passed by there last April they were making the logging road and taking in equipment for the workmen. .... .

Yours truly,
(Mrs) Ynes Mexia de Reygadas

P. S. Sometime ago I read an article that individuals were interesting themselves to save some very fine group of trees, whose location cannot recall, by getting different persons to buy each a tree. Can you tell me if anything ever came of this movement?

Letter from Newton B. Drury, Assistant Secretary, Save-the-Redwoods League, November 3, 1920.

My Dear Madam:

Mr. Sproul has asked me to acknowledge your letter of October 11th. Please pardon me for not having answered it sooner.

The Perfect Specimen

You will be glad to learn that the cutting in the heart of the Montgomery Grove has been stopped through the efforts of officials of the League and the citizens of Ukiah. This was accomplished by having a transfer of the holding upon which the owner was cutting for another holding on the outskirts of the Grove. There is a certain amount of cutting that this man is entitled to do on the edge of the Grove, but it will in no way impair the Grove's beauty ...

... Many thanks for your interesting suggestion about the saving of groups of trees by having individuals each purchase a tree. In effect this may be what will be done through the securing of donations from Individuals in order to save certain groups.

Yours very truly,
Newton B. Drury
Assistant Secretary

(Author's note: The League now encourages individual donations for trees and for groves.)

Letters from Ynes Mexia Collection, Save-the-Redwoods League, San Francisco, California.

Arequipa Sanatorium

Arequipa Sanatorium was founded by Dr. Brown. In his work with patients in San Francisco after the 1906 earthquake and fire, Dr. Brown discovered that the TB rate for women was twice that of men; appalled by this statistic, he made plans to build a sanatorium to treat women exclusively and called on many of his influential Bay Area friends to help.

A Marin County philanthropist donated land in Fairfax in western Marin County. This land had once belonged to Phoebe Apperson Hearst, a Brown family friend. The property adjoined Hill Farm, a home for convalescent women and children. An architect donated his services to design the sanatorium and Phoebe Hearst donated money for a laundry. An anonymous donor gave $10,000 so Arequipa Sanatorium could be opened in 1911. Arequipa is a Peruvian word meaning "Place of Rest."

Dr. Brown portrayed Arequipa as an enlightened health facility that would ameliorate disease and social dependency through a mixture of occupation and medical science. He described it as a place "where young women could go with their early tuberculosis and be cared for at a rate within their means with no element of charity and with the opportunity of earning part or all the cost through some form of work which they could do safely."

Conceived as a "school" where patients would learn how to cure themselves through fresh air and bed rest, the sanatorium featured large wards that were screened from floor to ceiling, even in winter. Whenever possible, locally grown food was served, and members of many Bay Area families donated money and goods. Arequipa eventually had three wards, a small library, living room, dining room, bathrooms, and examining rooms. Patients read, slept, wrote and published in-house magazines, and enjoyed the various entertainers who came to visit the sanatorium.

Dr. Brown believed that if the patients had something to occupy themselves, they would spend less time worrying about their disease and would heal more quickly. He began to experiment with various types of what is now called "occupational therapy," and in 1911 decided to open a pottery. Patients made pottery that was sold in stores throughout the country; profits helped pay the cost of their treatment. The operation closed at the end of World War I and Arequipa Pottery now is a prized collective today.

With the discovery of antibiotics in the 1940's and their use in the fight against TB in the 1950's, it became possible to treat patients at home, and admissions to the sanatorium dwindled. It was closed in 1957.

Information from Lilas Harley and Kathleen Barker Schwartz, "Philip King Brown and Areguipa Sanatorium: Early Occupational Therapy as Medical and Social Experiment," *American Journal of Occupational Therapy*, March/April 2013, Vol 67, e11-117, dol:10.5014/ajot.2013.005199.

THE PERFECT SPECIMEN

Sources

Note: All letters, unless otherwise noted, are from the Ynes Mexia Collection, Bancroft Library, University of California Berkeley.

Preface

p. 2, "I don't think there's ..." "U. C. Scientist Back from Trip into South America for Plants," *San Francisco News*, March 6, 1937.

Chapter One -- A Lonely Life

6, The information about Enrique Guillermo Antonio Mexia was obtained from Handbook of Texas Online, Teresa Palomo Acosta, "Mexia, Enrique Guillermo Antonio," accessed July 07, 2018, http://www.tshaonline.org/handbook/online/articles/fme74.

6-7, Information about the City of Mexia from http://cityofmexia.com/our-community/city-history.

8, "credited with installing ..." see Enrique Mexia bio above.

9 "To Papa: When ..." Part of ongoing letters from Yves Mexia to Enrique Mexia, January 14, 1882.

Chapter Two -- Adjusting to a New Life

12, Letters from Amanda Gray Mexia to Ynes Mexia regarding the health of Enrique Mexia, August 5-August 17, 1896.

p.13, Letter from Harris Etheridge and Knight, Attorneys, to Messrs. Kountze Bros., Bankers, New York City, July 26, 1899 regarding the legitimacy of the marriage between Mary Gray Mexia and Enrique Mexia.

17-19, Letters between Ynes Mexia and de Reygadas, November 20, 1910 to June 15, 1911.

19-21, Letter from Ynes Mexia to de Reygadas, July 31, 1911.

21, Sample of letters Ynes Mexia to de Reygadas, October 24, 1911 to December 29, 1911.

Chapter Three -- Finding a Niche
24-26, Dr. Philip Brown information from Lilas Harley and Kathleen Barker Schwartz, "Philip King Brown and Arequipa Sanatorium: Early Occupational Therapy as Medical and Social Experiment," *American Journal of Occupational Therapy*, March/April 2013, Vol 67e11e17.doi:10.5014/ajot.2013.005199.

27, "I do not think ..." Letter from Dr. Philip King Brown to Ynes Mexia, date unknown.

28, "To explore, enjoy ..." Sierra Club, "Sierra Club Purposes and Goals," http:www.sierraclub.org/policy,goals.asp

p. 29, "I am heartily ..." Ynes Mexia to Save-the-Redwoods League, November 3, 1919. Ynes Mexia Collection, Save-the-Redwoods League, San Francisco, CA.

29, "Cutting in the heart ..." Letter from Newton B. Drury, Assistant Secretary, Save-the-Redwoods League to Mrs. Ynes Mexia de Reygadas, November 3, 1920. Ynes Mexia Collection, Save-the-Redwoods League, San Francisco, California.

Chapter Four - Introduction to Botany
pp. 34-35, This story illustrates ... Letter from John Thomas Howell to Mrs. Doris Hollis Pemberton, June 20, 1980.

THE PERFECT SPECIMEN

36, "a good deal ..." and "There is no reason ..." Letter from Ynes Mexia to Alice Eastwood, July 25, 1925.

37, "I am willing ..." Letter from Ynes Mexia to Alice Eastwood, August 14, 1925.

Chapter Five -- Now on her Own
39, "Botany is indeed ..." A. G. Morton, *History of Botanical Science* (London: Academic Press, 1981), vi.

42-43, Botany is an ancient ... Ibid

Chapter Six -- Mexia's First Long Solo Expedition
44, "Transporting of equipment ..." Letter from Ynes Mexia to Dr. B. L. Robinson, July 2, 1926

45, "this gentleman knows . . " and "The streets when ..." Ynes Mexia, "Botanical Trails in Old Mexico -- the Lure of the Unknown," *Madono*, September 27, 1929, 227.

46, "it was hard ..." Ibid

46, "convolvulaceae of every ..." Mexia," Botanical Trails," 228.

47, "Many composites were" Ibid, 229.

48, "I privately think ..." and "keep up with ..." Letter from Ynes Mexia to unknown, October 24, 1926.

p.49, "Crops (and weeds) ..." Mexia, "Botanical Trails," 230.

49-50, "The mosquitos and ..." Mexia letter, October 24, 1926.

50, "The village was ..." Mexia, "Botanical Trails," 230.

50-51, "It was the ..." Letter from Ynes Mexia to unknown, November 11, 1926.

51, "The lagoons stretch ..." Ibid.

52, "It was eerie ..." Ibid.

Chapter Seven -- Into the Sierra Madre
53, "slept more comfortably ..." and "white toy houses ..." Letter from Ynes Mexia to Unknown, December 17, 1926.

54, "indubitable evidence ..." Ibid.

55, "like a silver ..." Ibid.

55-56, "...my driers were ..." Ibid.

56, "standing on edge," Mexia, "Botanical Trails," 233.

57, "Christmas dinner was ..."Letter from Ynes Mexia to Unknown, December 25, 1926.

pp. 57-58, "These are enormous ..." Letter from Ynes Mexia to Unknown, January 10, 1927.

58, "The varieties of ..." Mexia, "Botanical Trails," 233.

p. 59, "while the human population ..." and "Just behind the house ..." Mexia letter to Unknown, January 10, 1927.

p. 59, "Talk about the primitive! ..." Letter from Ynes Mexia to unknown, January 31, 1927.

60, "it is as clean ..." and "two or three ticklish ..." Ibid.

61, "Some very prickly ..." Ibid.

THE PERFECT SPECIMEN

62, "It is a great ..." and "We did not know ..." Letter from Ynes Mexia to unknown, February 28, 1927.

63, "It was a 'dip' ..." and "It grew black ..." Ibid.

64, "some of the most ..." and "fronds were four to ..." Ibid.

65, "just hugged the ..." and "This I firmly ..." and "I do not see ..." Ibid.

66, "Another like the ..." and "No wheel has ever..." Letter from Ynes Mexia to unknown, March 8, 1927

Chapter Eight -- Exploring Alaska, The Last Frontier

69, "most anxious to ..." Letter from Ynes Mexia to Dr. B. L. Robinson, September 27, 1928.

69-70, "My choice of ..." Letter from Ynes Mexia to Dr. B. L. Robinson, September 27, 1928.

70, "I could only ..." Letter from Ynes Mexia to Dr. Francis W. Pennell, December 10, 1928.

71, "a novel method ..." Ibid.

71, "made no impression ..." Letter from Ynes Mexia to Dr. Francis W. Pennell, September 30, 1929.

72, "only was able ..." Mexia letter to Robinson, September 27, 1928.

73, Mexia letter to Dr. Robinson, Ibid.

74, "I am mailing ..." Ibid.

p. 75, " I have had to ..." Letter from Ynes Mexia to Dr. Francis W. Pennell, December 10, 1928.

75, "I am very glad ..." Mexia letter to Pennell, September 30, 1929.

Chapter Nine -- Becoming Well Known

78, "To facilitate the ..." Letter from John Thomas Howell to Mrs. Doris Hollis Pemberton, June 20, 1980.

79-80, Bracelin information insert. Annetta M. Carter, "The Ynes Mexia Collections. and N. Floy (Mrs. H. P.) Bracelin," Notes and News, *Madrono*, Vol 23, 163-164.

81, An example of Mexia's Lectures.
Special Lecture Course No. 4, California Academy of Sciences announces a special course of free illustrated lectures on the general subject: The Beauties of Nature.
October 19 -- Up the Amazon and Over the Andes ... delivered by Ynes Mexia.

82, On June 10, 1927. Memorandum, Subject: 4 minute talk on Redwoods and the Save-the-Redwoods movement, Offfice of Save-the-Redwoods League, June 10, 1927.

pp. 82-83, "Mrs. Mexia was a close ..." Thomas letter to Pemberton, June 20, 1980.

Chapter Ten -- Exploring Brazil for a Year

84, "I have come up ..." Letter from Ynes Mexia to Agnes Chase, September 14, 1929.

85, "I unpack my ..." Letter from Ynes Mexia to unknown, October 15, 1929.

86, "The trip down ..." and "There was so much ..." and "An imp of perversity ..." Ibid.

The Perfect Specimen

87, "as any properly ..." and "I wanted to suggest ..." and "it has given ..." Ibid.

88, "perhaps the most ... ," and "had no idea ..." and "pure and simple ..." and "unpractical ..." and "we will get ..." and "or we would ..." Letter from Ynes Mexia to friends, November 22, 1929.

89, "green gloom," and "would have turned ..." and "green earth drapery," Ibid.

90. "stripped and pranced out ..." Ibid.

91, "She is an indefatigable ..." Ibid.

p. 91, "It is a fact ..." Letter from Ynes Mexia to Miss Pringle, December 16, 1929.

91-92, "Out of the ..." Ibid.

92, "This society has been ..." Letter from Ynes Mexia to unknown, January 23, 1930.

93, "The village was ..." Letter from Ynes Mexia to friends, February 3, 1930.

93, "In collecting ... ." Mexia letter to unknown, January 23, 1930.

94, "To be in a ..." Mexia letter to friends, February 3, 1930.

96, Insert on Brazilian cattle. "Glimpses of a Brazilian Cattle Ranch," (unpublished article, April 1931, Ynes Mexia Collection, Bancroft Library, University of California Berkeley, 4.

pp. 96-97, "One day I actually ... ." Ibid.

Chapter Eleven -- Up the Amazon

98, "Most of us ..." Ynes Mexia, "Three Thousand Miles up the Amazon," *Sierra Club Bulletin*, February 18, 1933.

100, "off the hoof" and "comfortable river steamer," Ibid

100, "The river itself ..." Ibid, 88-89.

101, "Manaos name was ..." "Following the Sun across South America or Up the Amazon and Over the Andes," (Unpublished article, October 30, 1933). Ynes Mexia Collection, Bancroft Library, University of California Berkeley.

103-104, "as the steamer ... ." Ibid.

104, "it was with feelings ... ." Ibid.

105, "Rather stunning they ..." Mexia, "Three Thousand Miles," 91.

105, "Iquitos is quite ..." Ibid, 92.

106, "freight, livestock and passengers ..." Mexia, "Following the Sun," op. cit.

106, "dumped ashore ..." Mexia, "Three Thousand Miles" op. cit.

107, "The shining, cream-brown ..." Ibid.

108, "One day, as ..." and "When they found ... ." Ibid, 93.

109, "a sort of ..." Ibid

109, "gloomy," Letter from Ynes Mexia to Dr. Bailey, April 5, 1932.

110, "The rainy season ..." Mexia, "Three Thousand Miles," 94.

111, "spewed us out," and "a lucky thrust," Ibid, 95.

Chapter Twelve -- Finding Ecuador's Wax Palm
113, "Ecuador, Land of ..." Ynes Mexia, "Camping on the Equator," *Sierra Club Bulletin*, February 22, 1937, 85.

115, "but the train sped ..." Letter from Ynes Mexia to My Dear Friends, September 26, 1934.

116, ". . .interminable reports ..." Ibid.

117, ". . . if these are the foothills and "rather small mule ..." Letter from Ynes Mexia to My Friends, November 21, 1934.

117, "I am continually surrounded ..." Letter from Ynes Mexia to Dear Far-Away Friends, February 13, 1935.

117-118, "Yet the farther I stray ..." Ibid.

118, ". .up and down ..." and "And I was foot sore ..." and "It is a beautiful ..." Ibid.

118-119, "Riding the oxen ..." and ". . .my Palomo was a treasure ..." Letter from Ynes Mexia to Dear Friends, April 7, 1935.

119-120, Mexia descriptions of getting the wax palm, Letter from Ynes Mexia to Dear Friends, July 15, 1935.

120, "He was right!" Ibid.

Chapter Thirteen -- South American Adventures Continue

p. 122, "I am not at all ..." and "She seemed a little ..." Letter from Nina Bracelin to Dr. Johnston, January 8, 1936, Ynes Mexia Collection, Bancroft Library, University of California, Berkeley.

p. 124, "still more beautiful ..." and "These were the original ... ." Ynes Mexia, "The Search for the Nicotianas," (lecture, California Academy of Sciences), 12.

125, "This, being an oceanic ..." Letter from Ynes Mexia to Bracie and Friends, March 12, 1936.

125, "After all Peru ..." Ynes Mexia to friends, June 1, 1936.

126, "a forlorn little port ..." and "The railroad is ..." and "A beautiful rose-red ..." Ibid.

126-127, "roofless but otherwise ..." Ibid, 3.

127, "It took the sun ..." Ibid.

p. 127, "The climate, scenery, soil ..." Letter from Ynes Mexia to Alice Eastwood, August 18, 1936.

Chapter Fourteen -- Final Exploration in Mexico

116, "Now that I'm back ... ." "U. C. Scientist Back from Trip into South America for Plants," *San Francisco News*, March 6, 1937.

pp. 131-132, "The old man, Severo, ..." Letter from Ynes Mexia to Nina Bracelin, November 15, 1937.

p. 132, "quick and smart ..." and "I stand the heat ..." and "probably it will be ..." and "personally I do not ..." Ibid.

The Perfect Specimen

133, "While I lost ..." Letter from Ynes Mexia to Alice Eastwood, February 3, 1938.

135, "I have been away ..." Ibid.

135, "They are very hospitable ..." and "Had to be reluctantly ..." Letter from Ynes Mexia to Dear Friends, March 8, 1938.

136, "glorious" and "where the tired mules ..." and "fine oak and spendid pines ..." Ibid.

Chapter Fifteen -- Leaving a Broad Legacy

137, "Fortunately, she had come ..." Letter from Mrs. H. P. Bracelin to Dr. A. C. Smith, July 26, 1938, Ynes Mexia Collection, Bancroft Library, University of California, Berkeley.

138, "like an older sister. . ." Ibid.

138, "This brave explorer and ..." "Memorials," *Sierra Club Bulletin*, June 1939.

pp, 138-139, "Resolved that the ..." "Resolution about Ynes Mexia," *Sierra Club Bulletin*, December 1938, xxix-xxx.

139, ". . .she collected approximately ..." Nina Bracelin, "Ynes Mexia," Memorials, *Science*, Vol 88, #2295, page 586.

139-140, Last Will and Testament of Ynes Mexia filed July 15, 1938 in City and County of San Fracisco, California.

140, "I have all of ..." Letter from Nina Bracelin to Ivan Johnston, August 6, 1938. Ynes Mexia Collection, Bancroft Library, University of California, Berkeley.

140, Last paragraph taken from a letter from John Thomas Howell to Mrs. Doris Hollis Pemberton, June 20, 1980.

141, "We understand that ..." Letter from California State Park and Recreation Commissioner Margaret Owings to Save-the-Redwoods League, January 20, 1968. Ynes Mexia Collection, Save-the-Redwoods League, San Francisco, California.

p. 142, "It is my pleasure ..." Letter from William Penn Mott, Jr., Secretary, California State Park and Recreation Commission, to Mr. Newton B. Drury, Secretary, Save-the-Redwoods League, January 25, 1968. Ynes Mexia Collection, Save-the-Redwoods League, San Francisco, California.

BIBLIOGRAPHY

Bartam, Edwin B. "Mosses of Western Mexico Collected by Mrs. Ynes Mexia." *Journal of Washington Academy of Science* (1928): 577-82

Bonta, Marcia Myers, ed. *American Women Afield: Writings by Pioneering Women Naturalists.* College Station: Texas A&M University Press, 1995.

Bracelin, Nina F. "Itinerary of Ynes Mexia in South America." *Madrono* 3 (1935): 174-76

Clark, Lewis F. "Mexia Resolution." *Sierra Club Bulletin*, December 1938.

Copeland, E. B. "Brazilian Ferns Collected by Ynes Mexia." *University of California Publications in Botany 17.* Berkeley: University of California Press, 1932.

"Enricque Guillermo Antonio Mexia." The Handbook of Texas Online, University of Texas at Austin. http://www.tsha,\.utexas.edu/handbook/online/articles/view/MM/fme74.html.

Goerke, Heinz. *Linnaeus.* Translated by Denver Lindley. New York: Charles Scribner's Sons, 1973.

Goodspeed, Thomas H. *Plant Hunters in the Andes.* Berkeley: University of California Press, 1961.

Goodspeed, Thomas H. and H. E. Stark. *University of California Publications in Botany 281.* Berkeley: University of California Press, 1955.

Durlynn Anema, Ph.D.

James, Edward T., ed. *Notable American Women, 1607-1950.* Cambridge, MA: Harvard University Press, 1971.

McLoone, Margo. *Women Explorers in North and South America.* Mankato, MN: Capstone Press, 1997.

Morton, A. G. *History of Botanical Science.* London: Academic Press, 1981.

National Park Service. "Denali." U. S. Department of the Interior. http://www.nps.gov/dena/.

Shearer, Benjamin F. and Barbara S. Shearer, eds. *Notable Women in the Life Sciences: A Biographical Dictionary.* Westport, CT: Greenwood Press, 1996.

Tyler-Whittle, Michael. *The Plant Hunters.* Philadelphia: Chilton Book Company, 1970.

"U. C. Scientist Back from Trip into South America for Plants." *San Francisco News*, March 6, 1937.

"Woman Braves Amazon Wilds for Specimens." *San Francisco Chronicle*, March 22, 1932.

Yount, Lisa. *A to Z of Women in Science and Math.* New York: Facts on File, Inc., 1999.

Mexia, Ynes. "Bird Study for Beginners." *Bird Lore* 27 (1925): 68-72, 137-141.

_____, "Birds of Brazil." *The Gull*, July/August 1930.

_____, "Botanical Trails of Old Mexico -- The Lure of the Unknown." *Madono 1* (September 27, 1929): 227-38.

_____, "Camping on the Equator." *Sierra Club Bulletin* 22 (February 1937): 85-91.

THE PERFECT SPECIMEN

_____, Collected Letters. Ynes Mexia Collection, Bancroft Library, University of California, Berkeley.

_____, "Down the San Pedro River." *The Gull*, December 1926.

_____, "Experiences in Hospitable Mexico." *Better Health*, October 1927, 432-58.

_____, "Following the Sun across South America or Up the Amazon and Over the Andes," (Unpublished article, October 30, 1933). Ynes Mexia Collection, Bancroft Library, University of California Berkeley.

_____, "Glimpses of a Brazilian Cattle Ranch." (unpublished, Ynes Mexia Collection, Bancroft Library, University of California, Berkeley, April 1931).

_____, "Ramphastidae." *The Gull*, July 1933.

_____, "Three Thousand Miles up the Amazon." *Sierra Club Bulletin 18* (February 1933): 88-96.

_____, "Vignettes of Birds Long Since Flown." *The Gull*, June 1935.

Web sites

http://www.botany.org/
The official Web site of the Botanical Society of America.

http://www.savetheredwoods.org/
The official Web site of the Save-the-Redwoods League.

http://www.sierraclub.org/
The official Web site of the Sierra Club.

INDEX

Academy of Natural Sciences: 67, 70, 72

Adams, Harriet Chalmers: 8, 42

Agricultural College (Vicosa, Brazil): 89, 92

Aguaruna Indians: 102-103, 104

Alacalufes Indians: 117

Alejandro: 116, 118

Arequipa Sanatorium: 140-141

Balsas River: 122

Bancroft Library, University of California, Berkeley: 130

Boyd, Louise Arner: 8

Bracelin, Nina Floy (Bracie): 72-76, 115, 120, 123, 127, 128, 129-130

Braga, Joaquin: 90-92

Britosh Museum of Natural History: 67

Brown, Philip King: 22, 27, 31, 75, 108, 127, 129, 140-141

California Academy of Sciences: 38-39, 77, 79, 80, 127

California Botanical Club: 37, 79

Chase, Agnes: 83, 86-88, 90, 92

Colby, William E.: 128, 129

Denali National Park and Preserve: 67

Drury, Newton B.: 33, 139-140

Dudley Herbarium (Stanford University): 39

Eastwood, Alice: 36-37, 40, 41 76, 80, 119, 123, 124-125

Eccleston, Samuel: 11

Ferris, Roxanna Stenchfield: 39-40

Field Museum of Natural History: 67

Ford, Henry: 98

Furlong, E. L.: 36

Goodspeed, Thomas H.: 15

Gray Herbarium (Harvard University): 68, 71, 72

Great Depression: 84

Harvard University Botanical Museum: 67

Hammeken, George Louis: 12

THE PERFECT SPECIMEN

Hopkins Marine Station:

Howell, John Thomas: 38-39, 77, 79, 80

Iahuas Indians: 100 ✓

Juarez, Benito: 12

Lake Titicaca: 116

Laue, Herman (first husband): 21

Machu Picchu: 118-119

Maranon River: 100-103, 105

Mexia, Adele (sister): 11, 14, 19-21, 122, 129

Mexia, Amanda Gray (half sister): 18-19, 129

Mexia, Clarence (half brother): 18-19

Mexia, Enrique (father): 11-15, 17

Mexia, Jose Antonio (grandfather): 12, 136-137

Mexia, Texas: 13

Mexia, Ynes
  Birth: 11
  Death: 128
  Death of first husband: 21
  Education: 14-15, 35-37, 39, 108, 120, 121-122
  Eugenia mexiae Standl: 55
  First marriage: 21

Mexianthus Mexicana: 131
Mimosa mexiae: 43
Publications: 79, 107, 121, 156-157
Second marriage: 22-26, 34

Montgomery Woods State Reserve: 32-33, 130, 131, 139

Morton, A.. G.: 44

mozos/guides: 46-47, 49, 54-56, 59-63, 101-105, 110, 111-113, 122-123, 124, 125

Mt. McKinley: 68, 69, 72

Muir, John: 22

New York Botanical Garden: 127

Niles, Mary Blair: 42

Ortega, J. Gonzalez: 46

Panama Canal: 84-85, 108, 120

Payne, Frances: 67-68

Peck, Annie: 42

Pennell, Francis W.: 68, 71, 92

Quichua Indians: 110

Reygadas, Augustin de (second husband): 22-26, 34, 137-138

Rio Negro: 98

Robinson, B. L.: 68, 71, 72

Royal Botanical Garden: 45

Santa Rosa: 108

Savage River: 69

Save-the-Redwoods League: 32-34, 75, 79-80, 108, 128, 129, 130, 139

Sierra Club: 31-32, 35, 75, 107, 108, 128, 129, 130

Sierra Madre: 53-54, 56-57

Smith, A. C.: 127-128

Smithsonian Institution: 83

Sproul, Robert G.: 139

Strait of Magellan

Tapajoz River: 98

Tierra del Fuego: 117

S. Department of Agriculture: 107. 115

University of California, Berkeley: 35, 76, 78, 108

University of Michigan, Ann Arbor: 121-122

Victoria: 97, 98

Wilmer, Sarah (mother): 11-12, 14

Wonder Lake: 69

Zinzer, Juan: 124-125